Chemical and B̶ cts
of
Electron-Spin oscopy

Chemical Aspects of Physical Techniques

Series Editor: Professor M. C. R. Symons
Department of Chemistry, University of Leicester

The aim of the books in this series is to cover the basic facts and theory and the applications of various physical techniques, from a point of view suitable for postgraduate and advanced undergraduate students of chemistry and related subjects. Books in the series are intended to show the reader what the techniques can do—what information they provide. The underlying physics will not be treated in depth, and the approach will not be highly mathematical.

Chemical and Biochemical Aspects of Electron-Spin Resonance Spectroscopy

Martyn Symons

Department of Chemistry
University of Leicester

 VAN NOSTRAND REINHOLD COMPANY

New York-Cincinnati-Toronto-London-Melbourne

ISBN: 0 442 30228 2 cloth
0 442 30229 0 paper

**Published by Van Nostrand Reinhold Company Ltd.,
Molly Millars Lane, Wokingham, Berkshire, England**

Van Nostrand Reinhold Australia Pty. Limited
17 Queen Street, Mitcham, Victoria 3132, Australia

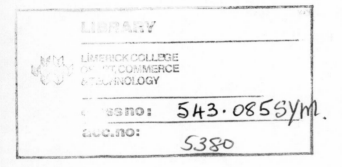
Text set in Northern Ireland at
The University Press (Belfast) Ltd.

Printed and bound at The Pitman Press, Bath

I dedicate this book to my beloved wife
Jan, and to the drivers of the Midland
Red buses on which it was written.

Preface

This book is about the ways in which the technique of e.s.r. spectroscopy has been used to probe certain areas of chemistry and biochemistry during the last 25 years. It is addressed to students whose desire is to 'see what e.s.r. can do' rather than to those who want to learn in depth about the underlying physics of this technique. It should be suitable for an undergraduate course, and for postgraduates and others who wish to use the technique. It was written as the result of many years of pressure from my own students.

Nevertheless, because of other pressures, it would never have been written had it not been for a decision made some 3 years ago to travel the ten miles between the village of Queniborough and Leicester University by Midland Red bus rather than by car. In fact, some 95% has been written during these journeys, and I therefore dedicate the book to the bus drivers, whose skill is such that my attention could be fully devoted to writing.

Many others have helped. My task has been greatly lightened by the helpfulness, skill and patience of Miss Vicky Orson-Wright and Mrs. Ann Crane who have typed from somewhat shaky manuscript and drawn from impressionistic diagrams, and never lost their sense of humour. I am also particularly endebted to Drs. P. W. Atkins, J. B. Raynor and Mr. J. A. Brivati for checking certain chapters.

Finally, I would not even have contemplated writing had it not been for the continuous underpinning of love and encouragement from my beloved wife, Jan.

Contents

Preface   vii

Chapter 1   1

1.1   Introduction   1
1.2   Scope and Limitations   3
1.3   Samples for e.s.r. Spectroscopy   4
1.4   Spectrometers   5
1.5   Spectra   6
1.6   Data   7
1.7   Intensities and Concentrations   8
1.8   Links with n.m.r.   9
1.9   Major Works on e.s.r. Spectroscopy   9

Chapter 2   The g-value   11

2.1   Basic Principles   11
2.2   The O_2^+ Radical   13
2.3   The O_2^- Radical   14
2.4   Links with Optical Spectra   15

Chapter 3   Hyperfine Coupling I   16

3.1   One Nucleus with $I = \frac{1}{2}$ 16
3.2   One Nucleus with $I > \frac{1}{2}$ 18
3.3   Several Nuclei   19
3.4   Isotropic Hyperfine Coupling   19
3.5   Spin Polarisation   20
3.6   Spin Polarisation for Atoms and Transition-Metal Ions   21
3.7   Hyperfine Coupling for the Methyl Radical   22
3.8   Calculation of Spin-Densities   23
3.9 π-Spin-Densities on Adjacent Atoms   23

Chapter 4   Hyperfine Coupling II   26

4.1   Electrons in p-Orbitals   26
4.2   Single Crystal Spectra   26

ix

4.3	Powder Spectra	28
4.4	Orbital Populations	30
4.5	Spin Polarisation	30
4.6	Anisotropy from d-Orbitals	31
4.7	Two Examples	31
	4.7.1 Example 1 (NO_3^{2-})	31
	4.7.2 Example 2 (NO_2^{2-})	31
4.8	Dipolar Coupling at Romote Nuclei	32

Chapter 5 Linewidths and Relaxation Effects 33

5.1	The Link between Widths and Relaxations: Line Shapes	33
5.2	T_1 and T_2	33
5.3	Chemical Exchange	36
5.4	Averaging Anisotropic Spectra	37
5.5	Rotations of Groups	38
5.6	Signs of Hyperfine Coupling Constants from Linewidths	39

Chapter 6 Examples from Organic Chemistry 41

6.1	Alkyl Radicals	41
6.2	β-Proton Hyperfine Coupling	43
6.3	Allyl, Vinyl and Related Radicals	46
6.4	Some Radicals Containing Nitrogen	48
6.5	Some Radicals Containing Oxygen	50
6.6	Nitroxide and Iminoxy Radicals	51
6.7	Spin-Trapping	51
6.8	Some Radicals Containing Fluorine	52
6.9	Aromatic Radicals	53

Chapter 7 Examples of Inorganic Radicals 58

7.1	Solvated and Trapped Electrons	58
7.2	Atoms	59
7.3	Diatomic Radicals	60
7.4	Triatomic Radicals (AB_2)	62
7.5	Tetraatomic Radicals (AB_3)	64
7.6	Pentaatomic Radicals (AB_4)	66
7.7	AB_5 and AB_6 Radicals	68
7.8	Cyclic Phosphazenes	69

Chapter 8 Some Organo-Inorganic Radicals 70

8.1	Carbon-Centred α-Radicals	70
8.2	Carbon-Centred β-Radicals	73
8.3	Sulphur-Centred Radicals and Related Species	77
8.4	Phosphorus-Centred Radicals and Related Species	79
8.5	Silicon-Centred Radicals and Related Species	80
8.6	Other Systems	81

Chapter 9   Environmental Effects   82

9.1   Ion-Pairs   82
       9.1.1   Radical Anions: Cation Hyperfine Coupling   82
       9.1.2   Anions with Two Binding Sites   84
       9.1.3   Solvation Versus Ion-Pair Formation   86
       9.1.4   Solvation of Ion-Pairs   86
       9.1.5   Triple Ions and Ion Clusters   88
       9.1.6   Radical Cations in Ion-Pairs   88
9.2   Solvation of Radical Anions   89
       9.2.1   Semiquinones   89
       9.2.2   Aromatic Nitro-Anions   90
9.3   Solvation of Neutral Radicals   90
9.4   Information from Radiation Studies   92

Chapter 10   Some Aspects of Mechanism   93

10.1   Generation of Radicals   94
        10.1.1   Photolysis   94
        10.1.2   Radiolysis   95
        10.1.3   Thermolysis   95
        10.1.4   Redox Processes   96
10.2   The Study of Radical Intermediates   97
        10.2.1   Additions and β-Eliminations   97
        10.2.2 α- and β-Scission in Phosphoranyl Radicals   99
        10.2.3   Electron Addition to Phosphorus(V)   100
        10.2.4   Dissociative Electron Capture   101
10.3   Kinetic Aspects   103
10.4   The Photosynthetic Process   106

Chapter 11   Polyelectron Systems   108

11.1   A Simple Model   108
11.2   Non-Axial Symmetry   111
11.3   Exchange Processes   112
11.4   Interactions Between Two Doublet-State Radicals (Di-radicals)   112
        11.4.1   Di-nitroxides   112
        11.4.2   Pair-Trapping   114
11.5   Ground-State Triplet Molecules   115
11.6   Photo-Excited Triplet-States   117
11.7   CIDNP and CIDEP   118

**Chapter 12   Transition-Metal Complexes Including those in Biological
                Systems**   123

12.1   The Use of Metal s and p Orbitals and Spin Polarisation   124
12.2   A Simple Model for $S = \frac{1}{2}$ Complexes   125
        12.2.1 d^1_{xy} Complexes   126
        12.2.2 $d^2_{xy} d^1_{xz}$ Complexes   126
        12.2.3 $d^2_{xy} d^2_{xz} d^1_{yz}$ Complexes   128
        12.2.4 $d^2_{xy} d^2_{xz} d^2_{yz} d^2_{z^2-y^2}$ Complexes   128
        12.2.5 $d^2_{xy} d^2_{xz} d^2_{yz} d^2_{x^2-y^2} d^2_{z^2}$ Complexes   128

12.3   Some Examples with $S = \frac{1}{2}$ 128
    12.3.1 d^1 Complexes   129
    12.3.2 Low-Spin d^5 (Ferric Haem Complexes)   132
    12.3.3 Low-Spin d^7 Complexes   134
    12.3.4 Vitamin B_{12} Derivatives   135
    12.3.5 Some d^9 Complexes   136
    12.3.6 Some Cu(II) Proteins and Related Systems   139
12.4   Complexes with $S > \frac{1}{2}$ 139
    12.4.1 $S = 1$ Complexes   140
    12.4.2 $S = \frac{3}{2}$ Complexes   140
    12.4.3 $S = \frac{5}{2}$ Complexes   142
    12.4.4 Met-Haemoglobin and Related Species   142
    12.4.5 Metal Ion Clusters   144
    12.4.6 Metal Ion Clusters in Biological Systems   147
12.5   Some Unusual Valence States   148
    12.5.1 Loss of Electrons from d^0 Complexes   149
    12.5.2 Gain of Electrons by d^{10} Complexes   149
    12.5.3 Complexes with Paramagnetic Ligands   150

Chapter 13 Some Biological Systems   152

13.1   Index of Biological Topics   152
13.2   Intrinsic Radical Signals from Biological Materials   152
13.3   Signals from Irradiated Materials   153
13.4   Spin-Labels   155

Appendix 1 Some Experimental Hints   158

A1.1 The Sample   158
A1.2 The Spectrometer   158
A1.3 The Spectrum   159

Appendix 2 Extracting Data   161

A2.1 Corrections   161
A2.2 Correction when A is Large ($S = \frac{1}{2}$)   162
    A2.2.1 Isotropic   162
    A2.2.2 Anisotropic   164
A2.3 Equivalent Nuclei when A is Large   165
    A2.3.1 Isotropic   165
    A2.3.2 Anisotropic Hyperfine Coupling   165
A2.4 Corrections when A is Small   167
A2.5 Nuclear Quadrupole Effects   170
A2.6 Corrections for Orbital Contributions to Hyperfine Coupling Constants   171
A2.7 Links with n.m.r. Spectroscopy   171

Appendix 3 ENDOR Spectroscopy   173

A3.1 The Experiment   173
A3.2 Advantages of ENDOR   173

**Appendix 4 Hyperfine Coupling Constants (G) for Unit Population of
Atomic Orbitals Calculated from Hartree-Fock Atomic
Wave-Functions**   176

References   178
Index   186

CHAPTER 1

1.1 Introduction

A molecule or ion is paramagnetic if it contains one or more unpaired electrons. These electrons will normally 'line up' along the direction (z) of an applied magnetic field (B_z), and this lifts the degeneracy of the $\pm\frac{1}{2}$ states of the unpaired electron (for $S = \frac{1}{2}$, or doublet-state species) as indicated in Fig. 1.1. Resonance is said to occur when the applied field induces a splitting whose energy is just equal to that of an applied microwave field. The spectroscopic act is then the flipping of the spin between the two quantised levels, $\pm\frac{1}{2}$, along the z direction as indicated in Fig. 1.2. It is useful to picture the electron in this field as describing a cone about z. The electron spin vector is pictured as precessing around the surface of this cone under the influence of the field, and can flip right over from the $+$ to the $-$ orientation when the precession rate equals the rate at which the magnetic component of the applied microwaves rotates.

In addition to having magnetism conferred by its intrinsic spin, an electron can acquire magnetism from its orbital motion (Chapter 2). This is also depicted in Fig. 1.2. The other major magnetism that we will be discussing arises from the nuclei (Chapters 3 and 4). The simplest magnetic nucleus is the proton, and its properties are also indicated in Fig. 1.2. It is important to note that the proton has the same spin angular momentum as the electron and so in zero-field these combine (Chapter 3). However, because of its far greater mass, its magnetic moment, μ_N, is very much smaller.

Electron spin resonance (e.s.r.) or electron paramagnetic resonance (e.p.r.) began just after World War II, more or less simultaneously in Russia and England [1.1,1.2]. The possibilities and scope for this form of spectroscopy had been envisaged by Van Vleck [1.3] who had, indeed, already developed a good deal of the underlying theories. The physicists who initially exploited this technique were solely concerned with transition-metal and lanthanide complexes, since these were the only common materials that were likely to give signals.

Chemists became more interested when it was demonstrated by various workers [1.4–1.6] that 'free radicals' could actually be detected and studied by this technique. This was an enormously exciting period and, for me at least, the excitement has not diminished during the subsequent twenty-four years. This is because the scope has constantly been widening, mainly because the instruments have steadily improved both in resolving power and in sensitivity. Whilst there is still, demonstrably, plenty to do in the fields of radical chemistry, triplet-state chemistry and transition-metal chemistry, nevertheless a great deal has now been accomplished. I think that the light that e.s.r. spectroscopy has shed upon electronic structure, reaction mechanism and solvation makes an

1

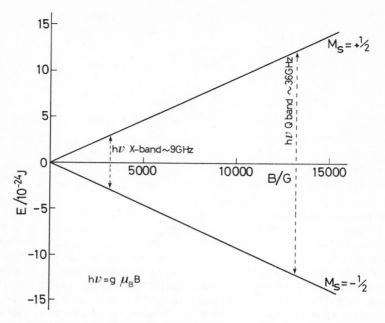

Figure 1.1 Effect of an applied, static, magnetic field, *B*, on the $M_S = \pm\frac{1}{2}$ levels of an electron with 'spin-only' magnetism. Transitions between these levels for the two commonly encountered spectrometers (X-band and Q-band) are shown by the vertical lines.

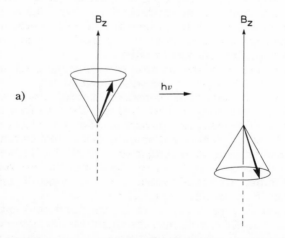

Figure 1.2 (a) The e.s.r. transition induced by the magnetic field in the microwave cavity. (b) Magnetic moments associated with (i) the electron spin angular momentum, S_z (ii) orbital angular momentum, L_z, and (iii) the proton spin angular momentum, $S(N)_z$. (μ_β is the Böhr magneton, the basic magnetic moment unit $= e\hbar/2m_e = 9.273 \times 10^{-28}$ JG^{-1} $= 0.273$ m^{2}A.) The proton has the *same* momentum as the electron, but μ_N is far smaller (5.051×10^{-31} JG^{-1}) than μ_β because of the far greater mass.

2

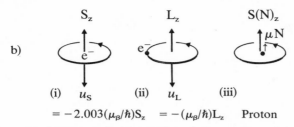

b)

S_z	L_z	$S(N)_z$
(i) u_S	(ii) u_L	(iii)
$= -2.003(\mu_B/\hbar)S_z$	$= -(\mu_B/\hbar)L_z$	Proton

outstandingly interesting story, and my aim in writing this book has been to endeavour to paint an overall picture which will, I hope, pass on this sense of pleasure and achievement to at least some of my readers. The picture is unashamedly impressionistic but, I hope, nevertheless worthwhile.

1.2 Scope and limitations

The major area of the subject that is totally neglected here is the rigorous mathematical development of the theory of magnetism and the undeniably complicated theory that underlies the various phenomena that are the subject of this book. I do this because I want to lead the reader as fast as possible into chemistry. I justify this approach on two major grounds: one is that all the major works on e.s.r. spectroscopy, listed at the end of this chapter, are primarily concerned with the underlying theory, and use the results primarily to illustrate the theory. My other justification is that in my own work, which has ranged widely across many fields of chemistry, I have rarely needed to delve into the underlying physics, but nevertheless my 'pictorial' views of what is happening do not seem to have let me down. Also, my experience of teaching e.s.r. spectroscopy tells me that students are quick to grasp the concepts without the need for the extensive use of algebra. There are, I am sure, many who would profit by understanding the chemical and biochemical revelations made by e.s.r. spectroscopy who do not wish to spend a long time on theory, and it is for them that this book has been written.

Two other areas have been omitted, mainly because of the need to keep the book a reasonable length. These are: paramagnetic lanthanide complexes and gas-phase spectroscopy. The former is important but highly specialised, and few chemists work in the area. Those who wish can learn all they need from Abragam and Bleaney [1.7]. The second area is also important, but is extremely limited since only a small number of molecules have been studied successfully. The topic is treated lucidly and in great depth by Carrington [1.8]. There are currently a number of exciting new techniques that are related both to e.s.r. and to microwave spectroscopy that promise to be of great use to gas-phase studies (see, for example, ref. 1.9).

The book is planned to give a minimum of the underlying principles in Chapters 2–5. Those already familiar with the technique are advised to avoid these chapters. However, in Chapter 6, attention is given to results in the organic field. Structure is stressed herein, and also interpretation. The way in which the radicals are formed, and mechanistic implications of their presence are considered subsequently, in Chapter 10.

In Chapter 7, a similar treatment is given for inorganic radicals. This chapter leans heavily on my earlier book on the subject written with Peter Atkins

3

[1.10]. A range of radicals fall somewhere between these two headings and these are put together in Chapter 8. Of course such classifications really are very arbitrary, but some sort of break-down is needed.

Before turning to the other two major areas, namely triplet-states and transition-metal complexes, some other aspects of the results for doublet-state ($S = \frac{1}{2}$) radicals are considered. In Chapter 9 I dwell on environmental factors. This is a favourite topic of mine, and of course e.s.r. spectroscopy is only one of many tools for studying solvation. Nevertheless, it has played an important rôle which is outlined herein. In Chapter 10 the very important mechanistic aspect of e.s.r. spectroscopy is outlined. My interest in 'free' radicals as intermediates in chemical reactions was initially aroused by reading the classic and beautifully written book by W. A. Waters, which is still well worth studying [1.11]. It is incredible just how much was understood before the technique of e.s.r. spectroscopy enabled chemists to detect the intermediates directly. All we managed to do, initially, was to confirm these concepts. However, in recent years much new insight of a remarkably detailed nature has been obtained.

In Chapter 11 aspects of triplet-state e.s.r. are touched on. This leads naturally to a discussion of transition-metal complexes in Chapter 12. This is a huge topic, and is currently extremely active. This is partly because chemists have at last realised how helpful the technique can be in shedding light on electronic structure in this field. Inevitably a single chapter cannot begin to do justice to such a wide field, so instead of treating the subject conventionally I have outlined the results that might be expected for certain simple cases, and hope it will at least show the reader the power and applicability of the technique.

Throughout, I have used examples of biological significance from time to time. I thought this was better than having a full chapter specifically oriented towards biology. Nevertheless, it is possible that some will wish to dwell primarily on these examples, so I have added a brief chapter (Chapter 13) that lists the topics covered and gives a range of references to others, together with a brief critical discussion of the utility of the method.

Finally, in three brief appendixes, I have listed some factual information that may be of use to those who actually plan to use the technique themselves. In Appendix 1 I have listed some arbitrarily selected 'cautions' and comments which may aid the beginner. In Appendix 2 I give a list of the adjustments that sometimes need to be made to raw data, especially if one wishes to use the results to obtain structural information. In Appendix 3 a very brief account of ENDOR (Electron-Nuclear Double Resonance) spectroscopy is presented. This form of double resonance has certain advantages and its use is mentioned from time to time in the main text. A full account can be found in ref. 1.12.

In the remainder of this chapter a few practical aspects are briefly outlined.

1.3 Samples for e.s.r. spectroscopy

These may be liquid or solid, and the latter may be in the form of glasses, fine powders or crystals. Most spectrometers work at X-band frequencies (*ca.* 9000 MHz) in which case about 0.2 cm³ of sample is sufficient. Remember that too big a sample may damp the cavity and hence lower the sensitivity. This is particularly true of fluid aqueous samples, which are best contained in capillary tubes or flat cells giving a relatively thin film of solution in the optimum region of the microwave cavity. Solutions require $<ca.$ 10^{-4} M of radicals in order to avoid spin-spin broadening (exchange broadening) and for high resolution studies dissolved oxygen should be removed because it is paramagnetic and hence can contribute to line-broadening. If you want to study frozen solutions, beware of phase separation, especially for aqueous systems. If this occurs, you will not be studying the system you wish to study. Also, paramagnetic solutes will aggregate and hence broad lines will result.

This is in fact my next point. The e.s.r. spectra of most pure paramagnetic solids comprise single broad lines. This is because the unpaired electrons are very close together and spin-exchange (Chapter 10) occurs. It may be possible to estimate g-values from single crystals of pure compounds, but the hyperfine splittings (Chapters 3 and 4) will be averaged to zero. Thus for single crystal work you need to grow diamagnetic crystals in which the paramagnetic species is dilutely incorporated. (For radiation studies, fortunately, the trapped radical centres are in fact formed dilutely throughout the crystal, so this limitation is automatically satisfied.)

1.4 Spectrometers

Today, many scientists use instruments of high complexity with complete efficacy, but without a full understanding of their inner workings. This book is

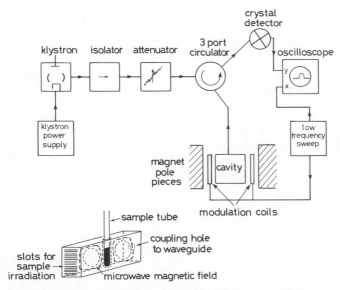

Figure 1.3 A simple e.s.r. spectrometer together with a normal X-band rectangular cavity. Electromagnetic radiation from the monochromatic klystron valve is led via wave-guides and normal microwave devices such as the isolator and attenuator to the 3 port circulator that feeds the energy to the microwave reflection cavity and to the crystal detector. In the cavity, stationary waves are set up such that the most intense region of magnetic field is at the sample, as illustrated. The large electromagnet provides a homogeneous magnetic field at the sample. This is swept by the power supply and is modulated at high frequency by the modulation coils, to give a phase-sensitive signal at the recorder.

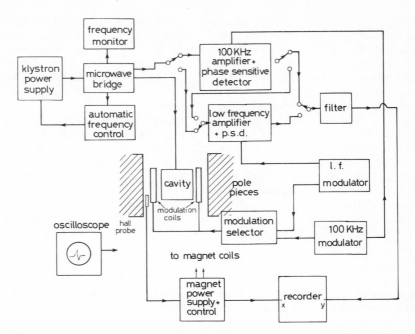

Figure 1.4 This gives a block diagram for a more advanced spectrometer. It is shown for readers with some experience in electronics, for their interest.

for people, especially students, who are concerned with the chemistry probed and revealed by e.s.r. spectroscopy rather than with the further development of complex instrumentation. Several excellent books are available in these areas. (See, for example, ref. 1.13.) What I have done is to include two diagrams; that in Fig. 1.3 gives the simplest plan for an elementary instrument of the type used in the early days. The other (Fig. 1.4) gives some details of a modern sophisticated instrument, although many refinements are still omitted. Those keen on instrumentation can, I hope, extract a considerable amount of detail therefrom.

1.5 Spectra

Typical single-line spectra are displayed in Fig. 1.5. Most modern instruments use a phase-sensitive method of detection, and hence produce the first derivative of the absorption spectrum. This is convenient since the peak position is then the base-line crossover, and the width is conveniently measured between the points of maximum slope, as indicated in the figure. If features are very poorly resolved, second derivative display makes them more apparent, and may help to convince you of their reality.

Sometimes it may be helpful to run difference spectra, in order to effect an instrumental 'subtraction'. This can conveniently be done using a dual cavity, with sample and 'reference' studied simultaneously. In general, 'markers', such as are needed in proton n.m.r. spectroscopy, are not needed, since instrument

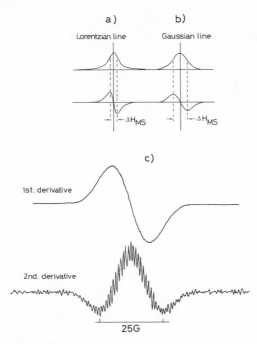

Figure 1.5 Line Shapes: (a) Lorentzian shape: Note weak but significant absorption in the wings. ΔH_{MS} is the symbol often used to indicate the width between the points of maximum slope. (b) Gaussian shape: broader at the peak and smaller extension in the wings. (c) First derivative of a line having a large number of components. Note the near Gaussian shape of the first derivative envelope, and the magnification of the weak undulations apparent in the second derivative, which resembles a normal absorption curve with the addition of negative wing features.

calibration is sufficiently accurate. However, workers often find it convenient to incorporate traces of reference materials which give narrow lines that act as a good check on the calibration.

1.6 Data

Most commercial instruments are calibrated to display spectra in Gauss (G). Most chemical spectroscopists still report their data, and presumably think, in terms of G, and certainly whenever I am involved in discussions about e.s.r. spectra it is assumed that the numbers refer to Gauss, unless one is told otherwise. This is surprising in view of the recommendation to use SI units, which have the Tesla (T) as the fundamental unit rather than G. Fortunately conversion only involves factors of $10^{-4}[G = 10^{-4}\,T]$, but very confusingly some spectroscopists use mT, so the numbers are just ten times smaller than in G. This *can* lead to confusion, since both numbers may be quite reasonable for a given system. Five years ago I would have used T not G, because I was convinced it would take over completely. Today I am confused, but after some thought have decided to use the unit used by the major manufacturers and the

7

Scheme 1.1

Conversions
$1\,G = 10^{-4}\,T$
$= 10^{-1}\,mT$
$= (g_{exp}/0.7145)\,MHz$
$= (g_{exp}/2.142 \times 10^4)\,cm^{-1}$

majority of chemists, namely G. Please, reader, don't be too critical—even more of you would object to T had I used that unit and, anyway, all you need to do is divide by 10 to get mT.

For some purposes it is necessary to convert G or T into frequency units before further information can be extracted from the data. Equations are given in Scheme 1.1. Most physicists report data in cm^{-1}, and frequently they use equations that extract the orbital magnetic contributions (see Appendix 2). I have used G or cm^{-1} throughout, for convenience.

Another problem that causes more heat than is really necessary relates to the symbols H and B. H is used to symbolise the magnetic field and its magnitude gives the field strength. B is the magnetic flux density or magnetic induction field. They used to have the same magnitude, but in SI units they differ by $4\pi \times 10^{-7}$. When we talk loosely about the magnitude of the field, we mean the flux density, and so I use B rather than H, to underline this point. However, H is widely used in the literature.

As explained in Chapters 2, 3 and 4, the parameters derived from the spectra are the g-values, the hyperfine coupling constants and, for $S > \frac{1}{2}$, the zero-field splitting parameters. These are all normally treated as tensor quantities, having three principal magnitudes and three principal directions associated with each. In solutions, rapid tumbling usually leads to complete averaging of the anisotropic part of these tensor quantities, leaving the isotropic part only (g_{av} and A_{iso}). Actually A and g are not strictly tensors, although they behave as if they were (see ref. 1.7). For convenience, and in common with most spectroscopists, we use the tensor terminology herein.

1.7 Intensities and concentrations

For n.m.r. spectra, the transition moment, and hence the area under the absorption curve, is a property of the nucleus and not of the molecule. So also, for e.s.r. spectra of $S = \frac{1}{2}$ systems, it is a property of the electron and not the radical, and again the area under the curve is directly proportional to the number of unpaired electrons. Unfortunately, it is not easy to use the instrument directly to determine this area, and hence spectrometers have to be calibrated using standard samples, such as the stable radical, $\alpha\alpha$-diphenyl β-picryl hydrazyl (d.p.p.h.). Even then, absolute accuracy is no better than ca. $\pm 20\%$.

Absorption curves can be integrated (twice for first-derivative spectra) by hand or electronically, but the most satisfactory method is to use a computer. Indeed, if there is appreciable noise in the spectra, a computer of average transients (CAT) can be an advantage, and quite accurate areas can be obtained. Care must be taken with Lorentzian curves not to truncate the lines, which have a large extension into the wings.

1.8 Links with n.m.r.

Since many students will be familiar with n.m.r. spectroscopy before reading this book, it is worth stressing the links that exist between these two techniques. Chemical shifts can be compared with g-value shifts, the 'marker' or standard in e.s.r. being the 'free' electron ($g = 2.0023$). In e.s.r. there is no vital need to use a shift marker since calibration is such that g-values can be estimated directly from a knowledge of the microwave frequency and the resonating field.

The hyperfine coupling can then be linked with nuclear spin-spin coupling. This is particularly true of the spin-polarisation mechanism (Chapter 3) since spin-spin coupling in the liquid phase is also largely transmitted by the bonding electrons.

Again, we can directly compare dipolar nuclear spin-spin coupling observed only in the solid state with dipolar electron spin-spin coupling discussed in Chapter 11.

Finally, we should ask the question, what does n.m.r. spectroscopy tell us about paramagnetic compounds? The answer has to be relatively little, in general, although in certain cases it can be extremely helpful. This is not the place for a detailed account of the problems involved (see, for example, ref. 1.14), but a few salient points can be noted. In general, short nuclear T_2 values (Chapter 5) in most paramagnetic systems result in such large linewidths that high-resolution n.m.r. is impossible. This stems from the contact interaction between nuclear and electron spins. If the electron spin can be made to invert very rapidly, the nuclei experience only the time-average effect, which depends on the population difference of the levels. This is found, naturally, for several transition-metal complexes, for which ligand nuclei give quite well resolved n.m.r. spectra, whose shifts now give information about their mode of interaction with the cations. Also, if concentrated solutions of radicals can be obtained, such that rapid electron transfer effectively inverts the electron spin experienced by a given nucleus, n.m.r. signals can again be observed. This is very useful since the direction of shift is dependent upon the sign of the hyperfine coupling constant, which can thus be determined.

1.9 Major works on e.s.r. spectroscopy

J. E. Wertz and J. R. Bolton, *Electron Spin Resonance*, McGraw-Hill, New York, 1972. (Excellent general text)

N. M. Atherton, *Electron Spin Resonance*, Halsted Press, London, 1973. (Excellent general text)

C. P. Poole, *Electron Spin Resonance: A Comprehensive Treatise on Experimental Techniques*, Wiley-Interscience, New York, 1967. (Mainly instrumental)

T. H. Wilmshurst, *Electron Spin Resonance Spectrometers*, Hilger, London, 1967. (Instrumental)

R. S. Alger, *Electron Paramagnetic Resonance Techniques and Applications,* Wiley-Interscience, New York, 1968. (Strong on the instrumental side)

A. Abragam and B. Bleaney, *Electron Paramagnetic Resonance of Transition Ions,* Oxford University Press, London, 1970. (Comprehensive on the theory of transition-metal complexes)

A. Carrington and A. D. McLachlan, *Introduction to Magnetic Resonance,* Hayser and Row, New York, 1962. (Covers the theory of n.m.r. and e.s.r. very helpfully)

L. A. Blumenfeld, V. V. Voevodski and A. G. Semenov, *Electron Spin Resonance in Chemistry,* Hilger, London, 1973. (Gives an insight into much Russian work)

P. W. Atkins and M. C. R. Symons, *The Structure of Inorganic Radicals,* Elsevier, Amsterdam, 1967. (Introduction to the technique and full details on the inorganic side)

CHAPTER 2

The g-value

2.1 Basic principles

The g-value in electron-spin resonance is the proportionality constant in the basic equation (for a system with $S = \frac{1}{2}$ and $I = 0$):

$$h\nu = g\mu_\beta B \qquad (2.1)$$

(for B in kG, ν in MHz, $g = 0.71446 \times \nu/B$) ν is the fixed frequency of the microwave radiation and B is the magnitude of the static field at resonance. It measures the rate of divergence of the $M_s = \pm\frac{1}{2}$ levels in a magnetic field, as depicted in Fig. 2.1.

If the electron spin is the only source of magnetism then $g_e = 2.0023$. (That $g_e \approx 2$ not 1 is explained by the Dirac equation and is related to the fact that the spin quantum number is $\frac{1}{2}$. The deviation from 2 can be explained by the theory of quantum electrodynamics.) For atoms and certain linear radicals, orbital angular momentum may be present in which case the orbital and spin magnetic moments combine together to give a total, J, and e.s.r. spectroscopy will monitor transitions between the J levels (see Fig. 1.2(b)). However, for most radicals, angular momentum is largely 'quenched' because the degeneracy of the orbitals involved is completely lifted by the covalent bonds. When the unpaired electron is in an orbital that is far removed from other levels, g will be close to the 'free-spin' or 'spin-only' value of 2.0023 (g_e). Fortunately g can usually be measured with great accuracy (usually $> \pm0.001$) and hence small deviations from 2.0023, which almost invariably occur, serve to help characterise the species and to separate one set of features from another.

The g-value is a unique property of the molecule as a whole and is independent of any electron-nuclear hyperfine interactions that may be present (*cf.* Chapter 3). In general, g is anisotropic, having three principal values along three orthogonal axes. (Thus g behaves like a second-rank tensor.) Deviations from the free-spin value arise because of the presence of orbital magnetism that adds to, or subtracts from, the spin magnetism. For most radicals this mixing is induced by the applied magnetic field, and is only a small perturbation on the spin magnetism, typical shifts being $\leq\pm0.01$. The extent of this mixing and hence the magnitude of the deviation from 2.0023 depends upon the magnitude of the spin-orbit coupling constants, ζ. Since ζ increases rapidly

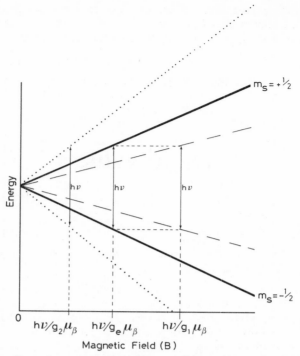

Figure 2.1 The effect of an applied static magnetic field (H) on the $m_s = \pm\frac{1}{2}$ levels of an unpaired electron. These are degenerate at zero field. The thick lines represent the spin-only behaviour, the transition ($h\nu$) occurring at a field corresponding to $g_e = 2.0023$. The dashed lines are for a radical having a low-lying vacant excited state, giving $g_1 < 2.0023$, and the dotted lines are for a radical having a neighbouring filled level, giving $g_2 > 2.0023$.

with atomic number, the shift, Δg, may be large if there is significant spin-density in p or d orbitals on heavy atoms in a molecule. Thus, when considering the form of the g-tensor for a given radical you can think of it as being made up of various separate contributions from spin-density on different atoms in the radical. For transition-metal complexes the central metal atom is usually the major source of g-shifts, but ligands comprising or containing heavy atoms may also make a significant contribution.

The magnetic moment for the electron arising from its orbital angular momentum is 1 rather than 2, so the g-value moves below 2 when the unpaired electron acquires orbital angular momentum (negative Δg). However, if coupling occurs via a filled rather than an empty level it is, in effect, a positive 'hole' that moves and hence the g-value shifts to values greater than 2 (positive Δg).

Often, the g-tensor is axial and one obtains $g_\parallel$ and $g_\perp$ from the spectra. However, in the general case there are three principal g-values (g_x, g_y and g_z). (*Note:* these are frequently given the symbols g_{xx}, g_{yy}, g_{zz} to underline the fact that they are along the principal axes of the g-tensor.) This is to be expected, since different levels are coupled by the magnetic field along these different directions. Powder spectra (in the absence of hyperfine coupling discussed in

12

Chapters 3 and 4) give these g-values directly, since spectral changes only occur at the turning points (see Fig. 2.2). For crystals, the transition moves between these limits, and in general follows the equation

$$g^2 = \alpha + \beta \cos 2\theta + \gamma \sin 2\theta \tag{2.2}$$

when θ is the rotation angle relative to the laboratory axes, and α, β and γ have to be determined from the sets of spectra obtained from the crystal. The usual procedure is to rotate the crystal about three orthogonal axes and measure the maximum and minimum g-values for each. Then, in each case,

$$\left. \begin{aligned} 2\alpha &= g_{max}^2 + g_{min}^2 \\ 2\beta &= (g_{max}^2 - g_{min}^2) \cos 2\theta_{max} \\ 2\gamma &= (g_{max}^2 - g_{min}^2) \sin 2\theta_{max} \end{aligned} \right\} \tag{2.3}$$

For certain linear radicals such as $\cdot OH$ or $\cdot O_2^-$, and for some paramagnetic transition-metal complexes, deviations from 2.0023 can be extremely large and the spin quantum number is no longer satisfactory. It is then necessary, as with atoms, to add the orbital and spin magnetic moments together and transitions between the resultant moment, J, must be considered.

Thus the g-tensor provides information about magnetically coupled excited states and the directional dependence of the coupling. This contrasts with the hyperfine interaction which is a property of the ground-state (Chapter 3). Detection of orbital magnetism can be very informative from a structural viewpoint. Two simple examples, O_2^+ and O_2^-, will serve to illustrate the factors involved.

2.2 The O_2^+ radical

The radical O_2^+ has been prepared in a variety of salts, such as $O_2^+SbF_6^-$. It has the electron configuration σ_1^2, σ_2^2, π_1^4, σ_3^2, π_2^1, with its unpaired electron in the otherwise empty antibonding π_2-orbital. In the absence of any crystal-field effects the real (x, y) forms of the π-orbitals are not appropriate, being replaced by the ± 1 orbitals with angular momentum about the z-axis. However, in the crystal lattice there are large electric fields that serve to define x and y, and to make π_x^* and π_y^* very different in energy (Fig. 2.2). Under these circumstances the unpaired electron can be thought of as π_x^{1*}, say, which will be the orbital that interacts least with the neighbouring anions.

Nevertheless, when the magnetic field direction is parallel to the molecular axis (z) it will induce orbital angular momentum and this will result in a shift to low g-values, i.e. to high-field. As indicated in Fig. 2.1, this shift increases linearly with field and hence is some 3.5× as great at Q-band (ca. 31.5 GHz) as it is at X-band (ca. 9 GHz). (This illustrates one of the major reasons for measuring e.s.r. spectra at two well removed frequencies. One can calculate accurately the relative shifts in peaks that are separated because of g-value differences and hence differentiate between these and hyperfine splittings (cf. Chapter 3). The same situation is found in n.m.r. spectroscopy: chemical shift differences are proportional to the magnetic field strength, but spin-spin couplings are not.)

There is a concomitant, but smaller, shift in $g_\perp$ to positive values (low-field). This arises for a variety of reasons and will not be analysed herein.

13

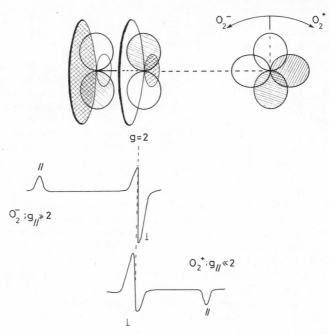

orbital momentum

Figure 2.2 π-orbitals for O_2^+ and O_2^-. For the former, e^- is in π_x^* and the other orbital, π_y^*, is imaginary. The crystal anions will lie preferably close to y. For O_2^-, e^- is again in π_x^*, and π_y^* is now full. Cations will now be preferentially close to y. Field along z induces e^- to move around z, thus causing a negative (O_2^+) or positive (O_2^-) shift in g_z.

2.3 The O_2^- radical

The situation for the well known superoxide ion, O_2^-, is reversed, since this has a single hole in the π^* level. Again we consider the case in which angular momentum is quenched, this time by neighbouring cations. The only way in which electrons can now circulate is by flow from π_x into π_y. This results in an induced magnetic field that augments the applied field, causing the resonance to shift to low-field and giving a correspondingly high g-value. Again, this is only induced by field along z ($\parallel$) (Fig. 2.2).

A variety of factors contribute to $g_\perp$. However, if $g_\parallel$ becomes large, orbital angular momentum about z can begin to control $g_\perp$ and this results in a shift to low g-values. Thus, a typical result for O_2^- is $g_\parallel = 2.175$ and $g_\perp = 1.999$, and these values are well reproduced by theory.

It is worth stressing that, since the orbital angular momentum that contributes to these g-shifts is a circular motion about the axis, transitions between the magnetically coupled states will not be induced by the electric vector of electromagnetic radiation and hence corresponding optical bands will be very weak. For transition-metal complexes these are typically the $d-d$ transitions that usually occur in or near the visible spectral region (Chapter 12).

14

2.4 Links with optical spectra

If the Δg values are small, perturbation theory can be used to link the g-shifts with suitable excitation energies, ΔE. Thus, for field along the principal direction x, we can write

$$\Delta g_x = f_x(C_A C_B \ldots, \lambda_A \lambda_B \ldots)/\Delta E_x \qquad (2.4)$$

where the ΔE_x values correspond to the energy gaps between the orbital of the unpaired electron and other nearby filled or empty orbitals that are linked to this by B_x. The $C_{A,B\ldots}$ values are the coefficients of the orbitals on atoms A, B, etc., and the $\lambda_{A,B\ldots}$ values are the appropriate spin-orbit coupling constants. If the orbital coefficients are known, or can be calculated, say from hyperfine coupling data, by using the atomic spin-orbit coupling constants, a value for the dominant ΔE term can be calculated. If available, this can be compared with optical data.

For unit spin-density in a p orbital Equation 2.4 simplifies to the equation given in Scheme 2.1. This also gives equations suitable for use in systems containing single unpaired electrons in d orbitals.

SCHEME 2.1

for ... p_z^1 $\Delta g_{\parallel} = 0$

$$\Delta g_{\perp} = \frac{\pm 2\lambda}{\Delta}$$

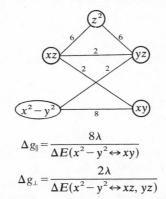

$\lambda = $ spin-orbit coupling constant

for ... d^1 systems

$$\Delta g = \pm \frac{n\lambda}{\Delta}$$

Use $\ominus$ sign, if coupling is to an empty level,

and $\oplus$ sign, if coupling is to a filled level.

The value of n which varies for different coupled levels is given by:

i.e. for $d_{x^2-y^2}^1 (d^9)$
$$\Delta g_{\parallel} = \frac{8\lambda}{\Delta E(x^2 - y^2 \leftrightarrow xy)}$$

$$\Delta g_{\perp} = \frac{2\lambda}{\Delta E(x^2 - y^2 \leftrightarrow xz, yz)}$$

15

CHAPTER 3

Hyperfine coupling I

Hyperfine coupling is the term used to describe the magnetic coupling that can occur between the spin of the unpaired electron (S) and those of nearby magnetic nuclei in the molecule (I). The phenomenon, which is undoubtedly the most important interaction in e.s.r. spectroscopy, resembles spin-spin coupling in n.m.r. spectroscopy. It is 'hyperfine' rather than 'fine' coupling for historical reasons—splitting arising from the coupling between two or more unpaired electrons being given the term 'fine structure'. In fact, both can be very small or very large.

3.1 One nucleus with $I = \frac{1}{2}$

In the high-field approximation, the interaction can be viewed in the simple manner depicted in Fig. 3.1, for a single nucleus with $I = \frac{1}{2}$. Clearly the states $M_s = +\frac{1}{2}$, $M_I = +\frac{1}{2}$; $M_s = +\frac{1}{2}$, $M_I = -\frac{1}{2}$ differ in energy. Given that the nuclear spin does not become reoriented during the electron spin transition, two lines will be detected in the e.s.r. spectrum. In fact, this intuitively obvious selection rule, $\Delta M_I = 0$, holds quite well in most circumstances (but not in all—see Appendix 2).

Before the application of an external field, the electron and nuclear spins are already coupled, and are properly treated in terms of their combined spins. This leads, for example, to apparent 'triplet' and 'singlet' states ($+1$, 0, -1) and 0 for a single $I = \frac{1}{2}$ nucleus, which differ slightly in energy, as indicated in Fig. 3.2. In the presence of an applied magnetic field, the ± 1 states diverge, but the 0 states are initially unaffected. As the field is increased the four states merge with the high-field states illustrated in Fig. 3.1. This 'zero-field splitting' has to be taken into account if the hyperfine interaction energy is comparable with the measuring frequency. When this is not the case, and the high-field approximation is valid, the simple equation

$$h\nu = g\mu_\beta (B + M_I A) \qquad (3.1)$$

holds satisfactorily. This defines A, the hyperfine coupling constant, in the manner in which it is most conveniently measured directly from the experimental spectra—as the separation between component peaks. Experimentalists frequently report their results using this definition, in units of Gauss (G) or in

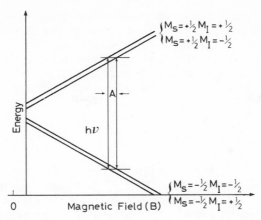

Figure 3.1 Divergence with field of the $M_s = \pm\frac{1}{2}$ levels in the presence of a single nucleus having $I = \frac{1}{2}$, in the high-field approximation. Note the two allowed transitions involve no change in M_I.

mT (Tesla) ($G = 10^{-4}$ T). In order to conform with usual chemical practice, raw data are reported in G herein. These become mT by dividing by 10. They can be converted into MHz using the relationship:

$$A \text{ (MHz)} = 2.8(g/g_e)A \text{ (G)} \tag{3.2}$$

or into cm^{-1}, using:

$$A \text{ (cm}^{-1}) = 0.935 \times 10^{-4} (g/g_e)A \text{ (G)} \tag{3.3}$$

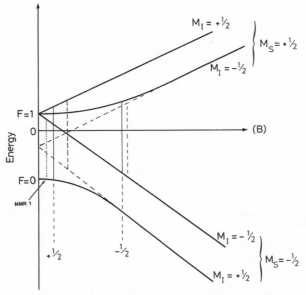

Figure 3.2 Energy levels for the same system as in Fig. 3.1 at low-field showing how the zero-field 'triplet' and 'singlet' arise. (The dotted transition labelled NMR 1 is not normally observed, see Section A2.2.1.)

17

The high-field approximation (Equation 3.1) is frequently invalid. The effect of the zero-field coupling is to move both absorption bands ($I = \frac{1}{2}$) to low-fields, the shift being greatest for the low-field component (see Fig. 3.2). Thus both the apparent hyperfine coupling and the g-value become too large. Allowance is made for this using the Breit–Rabi equation, which relates the field-values for the transitions to the real hyperfine and g-values. Some equations are given in Appendix 2.

3.2 One nucleus with $I > \frac{1}{2}$

In general, for $I = n$, there are $2n + 1$ energy levels and $2n + 1$ allowed transitions. So, for example for ^{14}N, $I = 1$ and there are three components, labelled +1, 0 and −1, in the e.s.r. spectrum for a radical containing a single coupled nitrogen nucleus. Similarly, ^{37}Cl, with $I = \frac{3}{2}$, gives rise to four allowed transitions, or ^{127}I, with $I = \frac{5}{2}$ gives six transitions. (These statements need to be modified if the hyperfine coupling is small and there is a large electric field gradient at the nucleus. This is because, for $I > \frac{1}{2}$, the nuclei are electrically asymmetric, having an electric quadrupole moment, and may tend to line up with the electric rather than the applied magnetic field.)

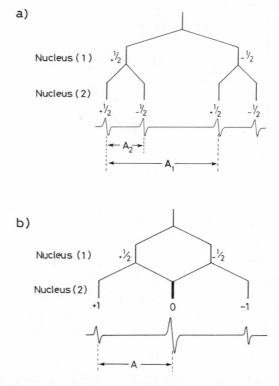

Figure 3.3 Evolution of e.s.r. spectra for radicals containing two nuclei with $I = \frac{1}{2}$ when they are (a) inequivalent and (b) equivalent.

18

3.3 Several nuclei

Two inequivalent nuclei with $I = \frac{1}{2}$ give rise to four allowed transitions $(M_I(1) = +\frac{1}{2}, M_I(2) = +\frac{1}{2})$, $(M_I(1) = +\frac{1}{2}, M_I(2) = -\frac{1}{2})$, $(M_I(1) = -\frac{1}{2}, M_I(2) = +\frac{1}{2})$ and $(M_I(1) = -\frac{1}{2}, M_I(2) = -\frac{1}{2})$. These are shown in Fig. 3.3, together with the case in which nuclei 1 and 2 are equivalent. In this case the two central lines coincide (in the high-field approximation only) to give a 1:2:1 pattern.

Many equivalent nuclei with $I = \frac{1}{2}$ give rise to multiplets with one line more than the number of nuclei, and a binomial distribution of intensities. Thus, for example, methyl gives four lines (1:3:3:1) and the benzene anion, $C_6H_6^-$, gives seven lines (1:6:15:20:15:6:1).

3.4 Isotropic hyperfine coupling

In most n.m.r. studies, solids are avoided because dipolar interactions between many magnetic nuclei broaden the features so greatly that much of the information is lost. These dipolar, through space, interactions average cleanly to zero and are lost in most liquid-phase spectra, and hence the liquid phase is overwhelmingly favoured. Fortunately, solid-state e.s.r. spectra are usually better defined and easier to interpret than their n.m.r. counterparts. The resulting anisotropic coupling constants are discussed in Chapter 4. In the liquid phase, these dipolar interactions are also averaged to zero and the spectra, which are in general more readily interpretable than those from the solid state, give the isotropic parameters. Many studies, especially in the field

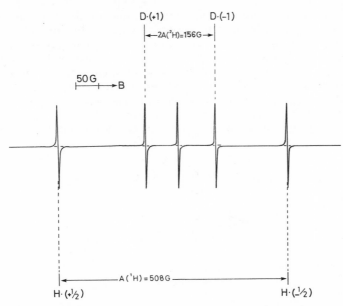

Figure 3.4 First derivative e.s.r. spectrum showing features for trapped hydrogen and deuterium atoms.

19

of organic radicals, are confined to the liquid phase, and our concern in this chapter is with the information that can be gleaned therefrom.

If attention is focused on the unpaired electron in a radical, one would expect that only spin-density in orbitals having s-character at the magnetic nuclei would give an isotropic interaction. (This is the 'Fermi contact' interaction. You can consider that the electron passes into the current loop that represents the source of the nuclear magnetic moment thereby experiencing a maximum possible magnetic interaction. This is only true for s-orbitals.) All other atomic orbitals have nodes at the nuclei and hence the direct isotropic interaction is zero. Arguments of this type are based upon the fact that, close to the individual nuclei in molecules, the orbitals resemble atomic orbitals. This is the reason why the use of parameters calculated for atoms is justifiable when discussing molecular parameters.

The simplest possible example of coupling arising from such direct population of s-orbitals is, of course, the hydrogen atom. These can be detected by e.s.r. spectroscopy in the gas phase, in liquids, and in solids. In all cases, spectra comprise narrow isotropic features, as shown in Fig. 3.4. This includes features for deuterium atoms in order to underline the large difference in hyperfine coupling, a difference which is frequently exploited in complex systems by specific deuteration. Note that the distance between the outer ($M_I = \pm 1$) lines for the deuterium atoms is equal to that for H· times the ratio of the magnetic moments. The ratio of the hyperfine coupling constants is *ca.* 6.5.

3.5 Spin polarisation

Since a large number of radicals have their unpaired electrons in π-orbitals, this leads us to the unpleasant conclusion that many radicals should not exhibit any hyperfine splittings in their liquid-phase spectra. Fortunately, as in n.m.r. spectroscopy, this is not the case. Splittings, albeit relatively small, are nearly always found. This is essentially because other electrons, mainly those involved in bonding, can be slightly 'spin polarised' so that, close to a given nucleus, one spin is preferred over another. The situation is depicted in Fig. 3.5 for the C—H bonding electrons of an $R_2\dot{C}H$ radical in which the unpaired electron is presumed to occupy a pure $2p$ orbital on carbon. The two bonding electrons are pictured close to the two nuclei since this is a dominant distribution in most

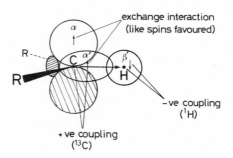

Figure 3.5 A qualitative representation of the way in which an unpaired electron in a $p(\pi)$ orbital on carbon influences the σ-electrons in a C—H bond by 'spin polarisation'.

20

σ-bonds. The electron close to carbon comes under the influence of the spin of the unpaired electron on this atom, and finds it slightly preferable to have the same spin. (This is an echo of Hund's rule of maximum multiplicity.) An important consequence is that the electron close to the hydrogen nucleus will have a bias in favour of the opposite spin. Since the σ-orbital looks like an sp^2 hybrid close to carbon, and a $1s$ orbital close to hydrogen, this has the effect of placing a small amount of spin-density in the s-orbitals and hence gives rise to some isotropic hyperfine coupling. The fact that the spin at the proton is opposite to that of the unpaired electron is accommodated by giving the resultant hyperfine coupling a negative sign. Only the magnitude of the coupling can be derived directly from the isotropic spectrum, but the sign can be determined in some cases from solid-state data, or from n.m.r. spectroscopy. The negative sign for coupling to such α-protons has in fact been verified.

Although this spin-polarisation effect is a great boon, it does tend to blur the distinction between pure π-orbitals and those with some direct s-character, and indeed inability to make this distinction has lead to considerable confusion, some of which still exists (*cf.* Chapter 6). However, theory and experiment show that unit spin-density on a given atom in a π-type orbital rarely, if ever, gives rise to more than *ca.* 4% spin polarisation induced coupling. (The 4% means that for unit spin-density on a given atom, A_{iso} corresponds to an apparent s-orbital population of 4%.)

3.6 Spin polarisation for atoms and transition-metal ions

For atoms such as N, O or F, having unpaired electrons in p-orbitals, it might be expected that A_{iso} would be zero, since there are now no valence orbitals with asymmetric radial extension to be polarised. You might suppose that any tendency to polarise a given electron to say, α-spin would be completely balanced by another with β-spin. This is very nearly true, but not quite. The extent of polarisation has fallen from *ca.* 2–4% to *ca.* 0.2%. One way of viewing the phenomenon is to think of one of the s electrons, say that with $+\frac{1}{2}$-spin, as occupying an orbital with a slightly greater radial extension than the other. The resulting small imbalance of spin-density at the nucleus causes the isotropic coupling.

This mechanism is also important for transition-metal atoms in complexes since their bonding may only be weakly covalent so that valence electron polarisation is relatively unimportant and inner s-shell polarisation may dominate. For d-electron systems the net effect usually turns out to be negative, so that A_{iso} is often negative. The magnitude of this contribution is usually *ca.* -0.4%.

We have seen that isotropic hyperfine coupling can be acquired by a direct occupation of atomic s-orbitals associated with the coupled nucleus or by an indirect polarisation of the electron spins of adjacent paired electrons. This gives a small, positive isotropic coupling if the electron is in a p-orbital on the nucleus concerned, and a small negative coupling if it is on an adjacent atom. The latter effect has been illustrated for an adjacent hydrogen atom (Fig. 3.5), but is quite general and needs to be taken into consideration for any adjacent atom or group. We illustrate this phenomenon by reference to the methyl radical.

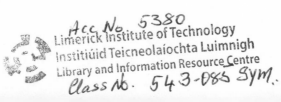

3.7 Hyperfine coupling for the methyl radical

A typical isotropic spectrum for $\cdot CH_3$ is shown in Fig. 3.6. This is for methyl radicals trapped in a rigid glass at 77 K, but these radicals are so small and interact so weakly with their environment that even at low temperatures they usually exhibit isotropic spectra. The important point to remember in understanding such a spectrum is that any one radical only contributes to a single absorption line. As discussed above, the lowest field line is from radicals that happen to have all three protons in the $M_I = -\frac{1}{2}$ orientation and the highest field line is from radicals having all $M_I = +\frac{1}{2}$ protons. The two inner lines correspond to two $-\frac{1}{2}$ and one $+\frac{1}{2}$ or two $+\frac{1}{2}$ and one $-\frac{1}{2}$ protons. There are three ways of doing this, and in first-order (for small coupling constants) these three components coincide to give two lines having three times the intensity of the two outer lines. These lines are usually labelled by the total M_I quantum number, as shown in Fig. 3.6. (If the isotropic coupling is large and the lines narrow, the inner lines from a set of equivalent nuclei may become resolved into sets. This phenomenon is discussed in Appendix 2.) Several pieces of information can be gleaned from this spectrum. Again, in first-order, the mid-point gives the g-value; the separation between adjacent lines gives the isotropic coupling in G (A_{iso}) (for small coupling constants the lines are equally separated); the $1:3:3:1$ intensity distribution and number of lines shows the presence of three equally coupled protons, and the integrated intensity gives the concentration. (Note that the spectrum is the first derivative of the

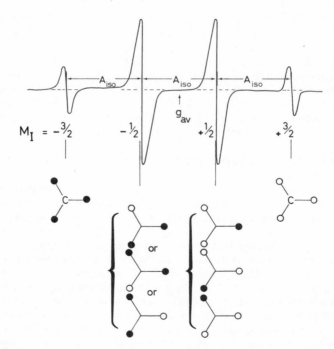

Figure 3.6 First-derivative e.s.r. spectrum for rotating methyl radicals.

absorption curve. As stressed in Chapter 1, most spectrometers use a phase-sensitive detecting system so that, unless deliberately changed, this is the way the spectrum emerges.)

At high gain, it is often possible to detect satellite features from $^{13}\cdot CH_3$ radicals. These would appear as replicas of the main quartet, *ca.* 20 G to high and low field, and with intensities *ca.* 1% of those of the central lines. The resulting isotropic coupling $A_{iso}(^{13}C)$, is 38.3 G. How can this be compared with the proton coupling of -23 G?

3.8 Calculation of spin-densities

For hydrogen atoms, the experimental hyperfine coupling is 508 G. Thus a rough measure of the spin-density on the protons in the methyl radical induced by spin polarisation can be obtained by dividing $(-)23$ G by 508 G (giving -0.045 or -4.5%). For carbon, there is no experimental value that can be invoked. However, there are theoretical values for the coupling that would be expected from unit population of s and p orbitals in atoms, and, at least for the lighter atoms, these are fairly reliable. (A table of values is given in Appendix 4.) Using the value for ^{13}C, we get a spin-density of 3.4% in the $2s$ orbital. If the $2p_x$ and $2p_y$ orbitals are included this gives a net polarisation of 10.2%. For each orbital this is 3.4%, which is quite close to the magnitude of the ^1H polarisation, as suggested by the crude model used.

This method of obtaining approximate spin-densities in different atomic orbitals on selected atoms in radicals will be used throughout. The method is purely utilitarian and useful for comparative purposes. It in no sense eliminates the need for full molecular orbital treatments, but nevertheless it receives support from the fact that in many cases, there is fair agreement from these values and those computed by the best available methods. In this example $(\cdot CH_3)$ the calculated spin-densities are indirect since the radical is planar. Obviously, for radicals in which the unpaired electron is in an orbital having direct s-character, the spin-density calculated from the isotropic coupling is a direct measure of this contribution.

3.9 π-spin-densities on adjacent atoms

Consider the two structures shown in Fig. 3.7. In (a) we consider an unpaired electron in a π^* orbital (i.e. NO, N_2^-, O_2^-, etc.), with some arbitrary distribution ρ_A and ρ_B in the two $p(\pi)$ atomic orbitals. (For ease of representation we use α and β instead of $\pm\frac{1}{2}$ to designate relative electron spins.) Spin on A (α_1) polarises other electrons on A, including those in the A—B σ-bond; giving α_1' on A and β_1' on B. Similarly α_2 spin on B gives α_2' on B and β_2' on A in the σ-bond. Thus the *net* polarisation of the σ-electron pair is $\alpha_1' + \beta_2'$ on A and $\alpha_2' + \beta_1'$ on B. This represents a major cancellation process that would reduce the polarisation in this bond to zero if $\alpha_1' \equiv \alpha_2'$, i.e. if $\rho_A = \rho_B$, as in N_2^-. Effective spin polarisation is then confined to other electrons on A and B, and in the absence of other σ-bonds this can be very small, as is the case for N_2^- (see Section 7.3). Obviously, provided both ρ_A and ρ_B are comparable, it is most important to consider these effects together. This is done, for the isotropic

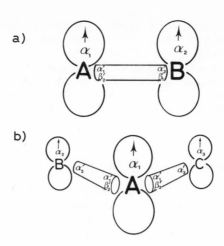

Figure 3.7 Acquisition of spin-density in s-orbitals by spin polarisation of σ-electrons (a) in the π-radical A—B spin-density α_1 on A gives α_1' and β_1' in the σ bond and α_2 on B gives α_2' and β_2'. (b) for the BAC radical with spin densities given by α_2 on B, α_1 on A and α_3 on C ($\alpha_2 + \alpha_1 + \alpha_3 = 1$). The net polarisation at A from the four σ electrons is $2\alpha_1' + \beta_2' + \beta_3'$.

hyperfine coupling, a, resulting from spin polarisation in π-radicals, by using the equations:

$$a_A = \rho_A Q^A + \rho_B Q^A_{BA} + \rho_C Q^A_{CA} + \ldots \tag{3.4}$$

$$a_B = \rho_B Q^B + \rho_A Q^A_{BA} + \ldots \tag{3.5}$$

Q^A, Q^B, etc., represent the total effect of polarisation from spin on A, B, etc., and the 'adjacent atom terms' Q^A_{BA}, etc., represent the specific effect of spin on any directly bonded neighbouring atoms, B, C, etc., on the A—B, A—C...σ bonding electrons. This is clearly zero for hydrogen, or fully bonded groups such as —CH_3, but can be significant whenever π-spin-density exists on B, C, etc. Since the Q values are specific to any nucleus, we have found it convenient and enlightening to convert these into U-values, by dividing by A^0, the calculated or measured hyperfine coupling constants for unit s-orbital population given in Appendix 4. Thus $U^A = 100 Q^A / A^{0,A}$ and $U^A_{BA} = 100 Q^A_{BA} / A^{0,A}$. The factor of 100 is included to give percentages [3.1].

The most striking result is that U^A values are largely confined in the narrow range of 2–4%, and this seems to reflect the number of σ-bonds involved. Thus $\dot{A}B_3$ radicals have values in the upper 3.5 to 4 range, $\dot{A}B_2$ radicals in the 2.5 to 3.5 range, and $\dot{A}B$ radicals in the 2–3 range. The consistently high values for $\dot{A}B_3$ radicals arise because all of the s orbital is involved in σ-bonding (as sp^2 hybrids for the symmetrical $\dot{A}B_3$ species). The more variable values for $\dot{A}B_2$ and especially $\dot{A}B$ radicals arise partially because s-character is now distributed between lone-pair orbitals and σ-orbitals, and generally the latter are the more attenuated and hence the more polarised.

The situation for the adjacent atom terms (i.e. Q^A_{BA}) is worse, since they depend crucially upon the s–p hybridisation of the σ bonding electrons at A. (I

24

stress that in this discussion we are concerned with liquid-phase spectra, and hence with isotropic hyperfine parameters: thus it is only s-spin-density that is involved. Since the s-contribution varies with structure, and with atomic number, Q_{BA}^A terms are very varied. We have presented a rough correlation that suggests values in the region of -1.0 for sp^3 or sp^2 orbitals. Clearly this falls off rapidly as the s-contribution falls below these commonly encountered values. In the special case of hydrogen as A, the s-contribution is unity and the value is in the region of -4.5%.)

In my experience, these generalisations are very helpful to give a quick understanding of the experimental results for inorganic and especially for . organic π-radicals. If there is not an approximate agreement then either the spin-density distribution is different from that predicted, or the radical is not truly π; that is, there may be some direct admixture of s-character into some of the orbitals.

Consider a few examples. For $^{13}CO_3^-$, $\rho_C \equiv 0$ and we expect sp^2 bonding at carbon. For $U_{OC}^0 = -1$ we calculate $a(^{13}C) = -11$ G, and the experimental value is $(-)11.2$ G. For NH_3^+ $\rho_N = 1.0$ and using $U^N = 3.5$ we calculate $a(^{14}N) = 19.25$ G. Experimental values range from 19.0 to 19.5 G. For $C_6H_6^-$, $\rho_C = \frac{1}{6}$ for all carbon atoms, hence

$$\frac{100a(^{13}C)}{A^0(^{13}C)} = \frac{1}{6}U^C + \frac{2}{6}U_{CC}^C$$

$$= \frac{1}{6}(3.5) + \frac{2}{6}(-1)$$

$$= +0.25$$

$$\therefore \quad a(^{13}C) = 2.8 \text{ G}$$

This is equal to the experimental value. Finally, a calculation that fails. Consider the nitroxide radicals, $R_2N\dot{O}$ that are so important in biological studies (Section 13.4). Solid-state studies show (from ^{14}N $2B$ term, see Chapter 4) that the spin-density distribution is close to 50% on oxygen and nitrogen. Hence, if the radicals are planar, we predict

$$\frac{100a_N}{A^0(^{14}N)} = 0.5 \times 3.5 + 0.5(-1) = 1.25$$

Hence $a(^{14}N)$ should be $ca.$ 7 G, but the experimental values range from 15 to 17 G. This suggests some slight admixture of $2s$ character into the orbital on nitrogen, and that is confirmed by X-ray diffraction studies on certain crystalline nitroxides.

CHAPTER 4

Hyperfine coupling II

4.1 Electrons in p-orbitals

In Chapter 3 we were concerned with isotropic hyperfine coupling, which is the only hyperfine information directly obtainable from radicals in solution. However, like the g-value, hyperfine interactions are usually anisotropic, with three principal values. Consider the form of the dipolar coupling expected for an electron in a pure p-orbital on a given atom with a nuclear spin $I = \frac{1}{2}$ (Fig. 4.1). It is easy to obtain this from a classical picture if one imagines the electron divided equally between two mean positions in the centres of the lobes. When the static external field is parallel to the axis of the p-orbital the interaction is as shown in Fig. 4.1(a), and is usually given the symbol $2B$. For field perpendicular to this axis, the field from the nucleus at the electron is reversed (4.1(b)). This is shown by a negative sign and since the field is now more dispersed, the value is reduced to $-B$. To this must be added the isotropic coupling, A, so that the values for $A_\parallel$ and $A_\perp$ become $(A + 2B)$ and $(A - B)$ respectively. Unfortunately, just measuring $A_\parallel$ and $A_\perp$ from the solid-state spectrum does not lead directly to A_{iso} and $2B$ because the signs of the coupling constants are not determined. However, if $|A_{iso}|$ is known from liquid-phase studies, the relative signs of $A_\parallel$ and $A_\perp$ are fixed. In many cases, they can be assigned unambiguously by reference to the necessary requirements of the radicals concerned. Examples are given in Chapters 6–8.

For a single crystal, suitably aligned, the doublet will change smoothly from a maximum separation of $A_\parallel$ (G) to a minimum separation of $A_\perp$ (G), but whatever the sign of $A_\perp$ the lines do not cross. Generally A_{iso} dominates the splitting, in which case the dipolar coupling follows the normal $(3 \cos^2 \theta - 1)$ law for dipolar coupling. However, when A_{iso} is small, this changes to $(3 \cos^2 \theta + 1)^{1/2}$, and hence the lines fail to cross [4.1]. This change arises because the anisotropic hyperfine field tries to retain control of the nucleus in competition with the applied field, and a resultant field needs to be considered.

4.2 Single crystal spectra

Ideally, single crystals are used to derive the anisotropic A- and g-values. For stable paramagnetic molecules or ions, the problem is to obtain some host

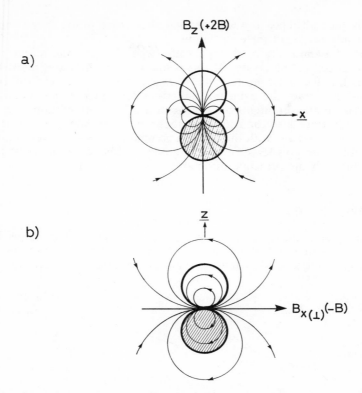

Figure 4.1 Field from a magnetic nucleus acting on a $2p$-electron: (a) for applied field (B) along the symmetry axis (Z) and (b) for a perpendicular field.

crystal in which the paramagnetic material is sufficiently soluble to give a crystal doped dilutely so that the individual paramagnets are well separated. (This is necessary to avoid spin-exchange or spin-spin broadening: Chapter 5.) Obviously, for transient species this is extremely difficult, but for species formed by high-energy radiation one can conveniently irradiate single crystals of the parent material. If the radiation is of sufficiently high energy (such as ^{60}Co γ-rays) it penetrates deeply into the crystal producing damage centres uniformly, which are usually well separated from each other. It is frequently found that these centres retain the original orientation of the parent molecules,

SCHEME 4.1

Derivation of A_{iso} and $2B$ and of orbital populations, a_s^2 and a_p^2	
$A_{\parallel} = A_{iso} + 2B$	$3A_{iso} = A_{\parallel} + 2A_{\perp}$
$A_{\perp} = A_{iso} - B$	
$a_s^2 = A_{iso}/A^{\circ}$	$a_p^2 = 2B/2B^{\circ}$

27

or possibly they all relax to a well defined alternative orientation—thus giving centres ideally arranged for e.s.r. study.

Even so, the simplicity of the spectrum depends upon the number of such sites per unit cell of the crystal. If this is one or two, the interpretative task is reaonable, but for a larger number it can become extremely difficult. The problem is to extract the principal directions and magnitudes of the g- and A-tensors from measurements on a range of arbitrary orientations. This can be done by collecting data for rotations about three orthogonal directions and diagonalising the results with the aid of a computer. It is, however, a great help to use results from powder spectra as a guide, and hence to arrive at rotations that pass through the principle directions.

4.3 Powder spectra

As we have seen, the g- and A-values can be expressed in terms of three principle values directed along three rectangular axes. Provided, as is commonly the case, the g- and A-tensors share the same axes, then powder spectra

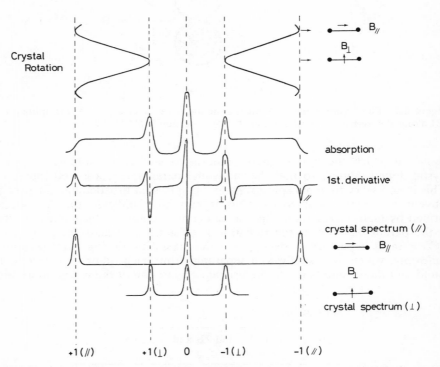

Figure 4.2 This figure illustrates the connection between a typical powder spectrum for a radical with a single coupled nucleus with $I = 1$, and the changes on rotation observed for a single crystal. The rotation is selected such that the parallel and perpendicular orientations are traversed. An isotropic g-value has been assumed: if $g_{\parallel} \neq g_{\perp}$, the central line would also fluctuate, and the ± 1 lines would sweep through different field ranges.

28

can give, directly, the principal values, but not, of course, the principal directions of these tensors. This is convenient, since the information is obtained far more directly and rapidly than from a crystal. Also, in many cases, frozen solutions and matrices containing reactive radicals can be obtained readily, when crystals containing them cannot.

A fine powder or glassy solution contains radicals statistically distributed over all orientations. However, no resonance is possible outside the limits of the field-values for the outermost components of the spectrum (usually the parallel features), so there is a sharp onset of absorption at these field positions. Similarly, one can expect to observe marked spectral changes at the other turning points. What is actually predicted and observed for axial and non-axial symmetry is summarised in the spectra given in Figs. 4.2 and 4.3.

Powder spectra can become extremely difficult to interpret when more than one type of magnetic nucleus contributes hyperfine features. Furthermore, under such circumstances it will often transpire that the different hyperfine tensors do not share principal directions, in which case the turning points will not, in general, give the principal values of these tensors. To obtain these with any accuracy, it is then necessary to study the radicals in single crystals, if this is possible. Nevertheless, the powder spectra remain extremely useful as 'fingerprints' for specific radicals since well defined features are usually present. In general, turning points or extrema which give powder features occur along the symmetry axes of the radicals, and in symmetry planes.

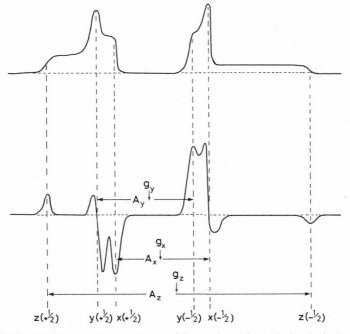

Figure 4.3 Predicted powder spectrum (absorption+first derivative) for a radical having non-axial symmetry and a single coupled nucleus with $I = \frac{1}{2}$. (*Note.* There are now several alternative ways of linking the features, but it is usually correct to link those which make the A-values as close to axial as possible, as shown.)

29

4.4 Orbital populations

Just as A_{iso} leads to an estimate of the s-character of an orbital near a given nucleus, so $2B$ can be used to give an estimate of p-character at this nucleus. Calculations for atoms give values for $2B°$ (see Appendix 4) which can be compared with the experimental value to give an approximate value for the spin-density in that orbital: (Scheme 4.1).

$$a_p^2 = 2B/2B° \tag{4.1}$$

4.5 Spin polarisation

We have already seen how paired electrons, especially those in bonding orbitals, can nevertheless contribute to the overall hyperfine coupling via the spin-polarisation mechanism. Obviously this will give a contribution to the total anisotropy in addition to the isotropic coupling discussed in Chapter 3. The latter is, however, the most important because a small admixture of s-character is much more noticeable than a comparable addition of p-character. For example, in ·CH_3, spin polarisation occurs to about 3.3% for the $2s$ orbital. If the same applies to $2p_x$ and $2p_y$ this will contribute a coupling of ca. -1.8 G along z. Allowance should be made for this contribution, but since it only reduces the measured direct contribution from e^- in $2p_z$ by ca. 3% it is usually ignored because of the many other uncertainties of comparable magnitude.

The effect, however, can be of significance when axial symmetry is absent and hence, for example, contributions from $2p_x$ and $2p_y$ differ. This difference can be measured and hence information regarding the symmetry of the centre is obtained.

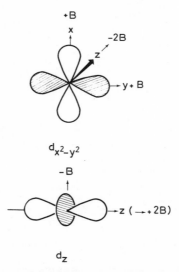

Figure 4.4 Dipolar hyperfine coupling for d-orbitals. For d_{z^2} the coupling resembles that to p-orbitals, with $A_{\parallel} = +2B$ and $A_{\perp} = -B$, for the remainder, the sign is reversed, with $A_{\parallel} = -2B$ and $A_{\perp} = +B$.

4.6 Anisotropy from d-orbitals

The situation is similar to that for p-orbitals, but the change in orbital symmetry has the effect of changing the sign of the interaction as shown in Fig. 4.4, except in the special case of d_{z^2}. The classical dipolar model again accounts well for the form of the coupling.

Once again, from the magnitude of the measured anisotropy and that calculated for unit occupancy, an estimate of the orbital population can be derived. If the sign of the anisotropic coupling can be deduced, as is often the case, then a choice between an orbital of d_{z^2} symmetry and one of the remainder can be made. If single crystals of known structure can be used, the directions of the hyperfine components fix the magnetic axes so that the precise d-orbital involved is established.

4.7 Two examples

Since the information obtained from hyperfine coupling constants is extremely useful in the study of radicals, two examples are given.

4.7.1 *Example 1*

This radical has a $1:1:1$ triplet spectrum suggesting hyperfine coupling to ^{14}N, with $A_\parallel = \pm 61.5$ G and $A_\perp = \pm 30.5$ G. If both signs are positive, we get $A_{iso} = 40.83$ G and $2B = 20.67$ G, but if $A_\perp$ is negative, we get $A_{iso} = 0.17$ G and $2B = 61.33$ G, using the equations in Scheme 4.1. If we assume that the nucleus is ^{14}N, then $a_s^2 = 0.074$ and $a_p^2 = 0.61$, or $a_s^2 = 0.0003$ and $a_p^2 = 1.8$. The latter is clearly rejectable, and the other sign combinations are also rejectable, so that $A_\perp$ must be positive, and the orbital populations are established.

(*Note.* This radical, formed in KCl crystals doped with nitrate ions, has been identified as $\cdot NO_3^{2-}$. The data show that this ion must be pyramidal.)

4.7.2 *Example 2*

This radical also gave a hyperfine triplet and hence ^{14}N is again suspected. With $A_\parallel = 30.9$ G, and $A_\perp$ slightly split to give $A_x = \pm 5.0$ G and $A_y = \pm 7.0$ G, we have $A_{iso} = +14.3$ G and $2B \approx +16.6$ G or $A_{iso} = +6.3$ G and $2B \approx +24.6$ G, depending on the sign of '$A_\perp$'.

(*Note.* We again assume that $A_\parallel$ is positive, since a negative value would give impossible results: also, we ignore the difference between A_x and A_y and use the average of 6 G.) If the nucleus is ^{14}N, then the first set gives $a_s^2 = 0.026$ and $a_p^2 = 0.49$ whilst the second gives $a_s^2 = 0.011$ and $a_p^2 = 0.72$. Either result is reasonable, and without further information we cannot make a definitive selection. Note that if we knew A_{iso} from a liquid-phase experiment, or from a matrix study in which the radical was free to rotate, a definitive choice could be made. In this case, the radical was formed from NO_2^- in potassium chloride crystals, and is thought to be NO_2^{2-}. Comparison of the data with those for similar radicals strongly favours the positive sign for $A_\perp$.

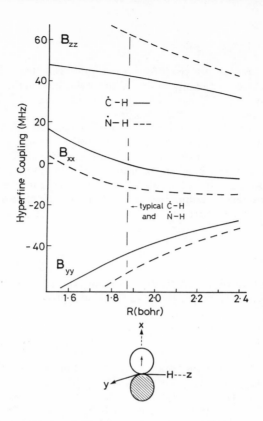

Figure 4.5 The way in which the hyperfine coupling (in MHz) to ^{1}H in planar $R_2\dot{C}$—H or $R_2\overset{\cdot}{N}H^+$ radicals changes with R, the X—H bond length.

4.8 Dipolar coupling at remote nuclei

So far, we have only considered dipolar coupling caused by electron spin in p or d orbitals on the nucleus at the centre of these orbitals. Since this form of coupling is proportional to $\langle r^{-3} \rangle$, it is usually insignificant at more remote nuclei. However, in the particular case of α-protons it is certainly not insignificant. The situation is depicted in Fig. 4.5 as a function of bond length for $>$CH and $>$NH radicals, assuming unit spin-density on carbon or nitrogen. Experimental values fall close to the vertical dashed line. Note that, for $>$C—H radicals, B_x is fortuitously close to zero so the anisotropic coupling takes the form $|+B, 0, -B|$ along the z, x and y axes. Typical experimental values, adding in A_{iso} at, say, -22 G, would be $|-9, -22, -35|$ G, for such radicals.

CHAPTER 5

Linewidths and relaxation effects

5.1 The link between widths and relaxations: Line shapes

There are two major contributions to the widths of e.s.r. absorptions: unresolved hyperfine splittings, and kinetic processes. The former can give rise to a whole range of contours, depending upon the numbers of lines involved, their separation and their relative intensities. This leads to a Gaussian shape in limiting cases, an example being shown in Fig. 1.5(c). By varying the conditions, part or all of such concealed splitting can sometimes be resolved. If not, then 'resolution' may be achievable by using ENDOR (Appendix 3).

Our concern in this chapter is with kinetic broadening effects, which, if dominant, result in a Lorentzian line shape (*cf.* Fig. 1.5).

5.2 T_1 and T_2

Two important mechanisms for line-broadening need to be considered. One, involving actual electron spin reorientations and hence a change in magnetisation, is called 'nonsecular' and given the label T_1. The other, which can be thought of as stemming from fluctuations in the energies of the ground and excited states or in the local field experienced by the electrons, is a 'secular' process, labelled T_2 (Fig. 5.1). A third process, involving the reorientation of nuclear spins, is termed 'pseudosecular'. This process, which also contributes to T_2, is generally of minor importance, except in certain exchange processes. The special case of spin-exchange between radicals in relatively high concentration is considered in Chapter 9.

The T_1 process controls the time taken for a system to attain thermal equilibrium in a magnetic field. If T_1 is long, relaxation is inefficient and the system will 'saturate' readily. Low microwave powers must then be used to avoid loss of signal. If T_1 is short, high powers can be used, but there may be a life-time uncertainty contribution to the width. As with n.m.r. spectroscopy, T_1 can be obtained by, for example, direct measurements on the rate of gain of signal after the system has been subjected to an intense pulse of power. This rate gives $1/T_1$ directly, but special equipment is needed, and such measurements are rare.

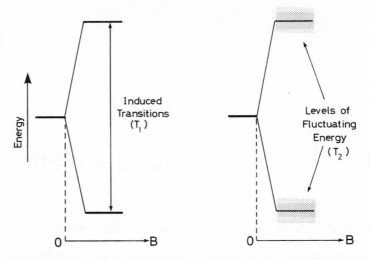

Figure 5.1 T_1 (nonsecular: spin-lattice) and T_2 (secular, spin-spin) processes.

In order to cause a spin transition, there has to be a local fluctuating magnetic field having just the right frequency at the electron concerned. For dilute solutions this is clearly not highly probable. If the resonance is broad, the probability of spin inversion increases. Thus T_2 processes that increase the linewidth also increase the probability of inversion. If there is a large orbital magnetism (Δg large) this is coupled to electrical fluctuations, and the resulting magnetic fluctuations can cause spin-inversion.

In contrast with T_1, any local field will contribute to T_2 by modifying the instantaneous local field at the electron (Fig. 5.1). If we could look at a single

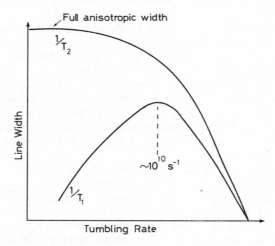

Figure 5.2 Variation in line width with the rate at which the radicals tumble in solution. For anisotropic species the contribution from T_2 clearly dominates except at high rates.

34

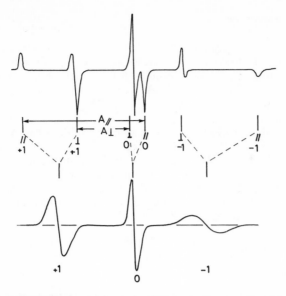

Figure 5.3 This hypothetical set of spectra illustrate the way in which the anisotropic powder spectrum 'controls' the form of the solution spectrum for a radical with a single coupled nucleus with $I = 1$. Note that the solution lines are isotropic, but their width is proportional to the field through which the lines 'move' as the radicals tumble. If other factors can be allowed for, this differential broadening leads to a measure of T_c, the rate at which the molecules tumble.

radical, and if fluctuations in local fields were sufficiently slow, we would observe a resonance that was continuously and randomly moving over a range of field. Obviously the other signals from the ensemble of electrons would be moving quite differently. When these fluctuations occur rapidly, as is the rule in the liquid phase, all the electrons resonate at a time-averaged position, so the line is far narrower than would have been the case in the slow fluctuation limit (Fig. 5.2). This averaging is the same phenomenon that makes n.m.r. lines narrow in the liquid phase despite their considerable widths in the solid state. It is also the same as that which causes a signal comprising two or three principle g-values to average to a single line in the liquid phase, and anisotropic hyperfine couplings to appear at the average (isotropic) values. The way in which this happens for a hypothetical ensemble of radicals having g- and A-anisotropy is shown in Fig. 5.3. In this case, the central ($M_I = 0$) line is narrowest, because ΔB_0 is small. The $M_I = -1$ line is the broadest because ΔB_{-1} is at a maximum.

Averaging effects of this type sometimes present conceptual difficulties. A crude picture of such processes can be obtained by recalling the way waves on the surface of water affect a floating log. When the motion is of low frequency the log will tend to follow and will tilt and move as the waves move. However, higher frequency 'ripples' lap against the sides of the log but cause no resultant motion at all. In other words, the log remains in its averaged (horizontal) position.

5.3 Chemical exchange

This averaging over a range of possible resonances has much in common with the situation that occurs when, say, two precisely positioned resonances are made to interchange (Fig. 5.4). Consider, for example, a base-catalysed proton-transfer process, in which $[\cdot AH] > [\cdot A^-]$,

$$\cdot AH + B^- \rightleftharpoons \cdot A^- + BH \tag{5.1}$$

Every time the radical $\cdot AH$ loses and regains its acidic proton, there is an even probability that the nuclear spin of the proton will be inverted. In 'slow exchange' this will cause a broadening of the doublet features associated with $\cdot AH$, but as the rate increases the doublet is lost and a broad central line appears, which becomes narrower as the rate increases further. This is called 'fast exchange'. In the slow-exchange situation, the broadening depends only upon the rate of exchange:

$$\delta \propto \omega \tag{5.2}$$

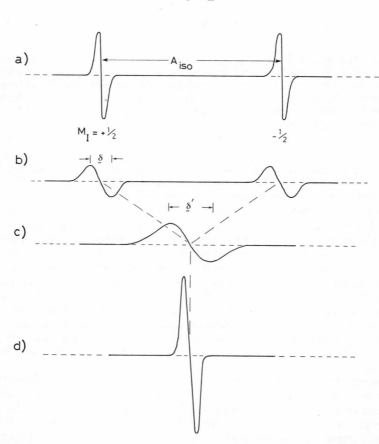

Figure 5.4 Hyperfine features for a radical $\cdot AH$ undergoing proton exchange (Equation 5.1). (a) No exchange. (b) Slow exchange. (c) Fast exchange. (d) Very fast exchange.

when ω is the 'jump frequency'. However, for fast exchange the broadening also depends upon the separation between the lines $A(^1H)$:

$$\delta' \propto \frac{A^2(^1H)}{\omega} \tag{5.3}$$

In this particular example, we are really considering linewidth control by an effective inversion of the proton; that is, in effect, a pseudosecular process. The same result would be obtained if, for some reason, the protons were made to invert without any actual exchange. Exchanges involving other nuclei, notably those of alkali-metal cations in ion-pairs, give similar results. Also exchange of electrons can be an important method for loss of hyperfine coupling. Examples are given in Chapters 9 and 10.

5.4 Averaging anisotropic spectra

Radicals with anisotropic g- and A-values, tumbling in solution, are usually in the fast-exchange situation. Isotropic lines are then obtained, which become narrower as the tumbling rate increases (for example on heating or lowering the local viscosity). It is clear from Fig. 5.3 that each line will be affected differently, since for fast averaging the width depends upon ΔB^2 (equation 5.3), where in this case ΔB can be thought of as the total spread of the resonance for that line. This is indeed found for radicals or transition-metal complexes at low temperatures and for viscous solutions. The 'slow-exchange' situation is more difficult to envisage, because a complete range of possible line positions are involved instead of a small number. What one observes is a spectrum resembling that of the rigid solid, but with the apparent turning points (often parallel and perpendicular features) shifted towards the average position. When motion is so slow that this apparently partial averaging occurs, it is important to

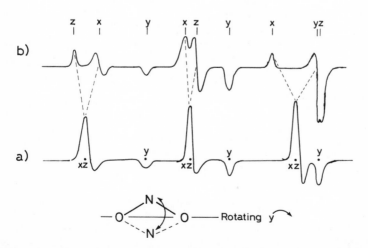

Figure 5.5 First derivative e.s.r. spectrum for NO_2 radicals (a) in a matrix in which rotation about the long axis (y) is rapid and (b) in a matrix that inhibits all rotations.

consider the possibility that motion about different axes occurs at different rates. This can be extremely important for radicals in solids, when motion may be severely restricted about one axis, but relatively free about another. This may be controlled by the shape of the cavities containing the radicals, or by the presence of weak bonding to the lattice. An example, that of NO_2 rotating about its long axis, is shown in Fig. 5.5. Failure to recognise this mode of rotation led to some confusion in early studies of this radical in rare-gas matrixes.

Very slow tumbling is also a feature of radicals that are an integral part of large biopolymer molecules whose tumbling rates are usually slow on the e.s.r. time-scale (Section 13.4).

5.5 Rotations of groups

Another important form of motional averaging arises when specific groups in a radical undergo rotation (or pseudo-rotation). One example already cited is for

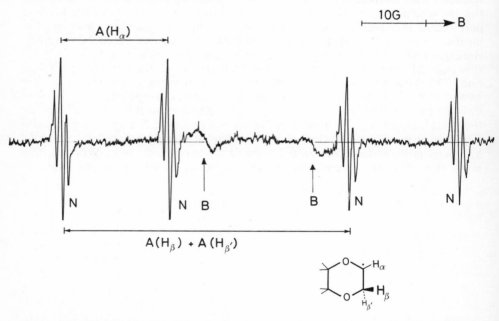

Figure 5.6 First derivative X-band e.s.r. spectrum for an aqueous solution containing the radical

Note that the third and fourth lines are very broad relative to the remainder. On heating, these broad features would ultimately resemble those of the other components, but with double the intensity. Note the weak coupling to two extra protons, from the CH_2 group of the unit $—CH_2O—\dot{C}H$.

38

a methyl group in an alkyl radical, for example in ethyl, $H_2\dot{C}$—CH_3. As discussed in Chapter 6, if there were no rotation about the C—C bond, the three methyl protons would in general give distinct and different hyperfine couplings. However, if relatively rapid rotation about the C—C bond occurs, four isotropic lines appear at averaged positions.

The situation becomes more complicated for asymmetric groups as, for example, in the propyl radical, $H_2\dot{C}CH_2CH_3$. There is now a preference for a particular conformation at low temperatures, whose structure can be deduced from the magnitude of the β-proton hyperfine coupling constants. This preference is usually small, and as the temperature is raised librations about the preferred conformation set in, thus modifying the proton coupling which move towards the average value, with appropriate line-broadening. Ultimately, the barriers are surmounted and 'free' rotation sets in, giving the average coupling constants. The preferred conformation may be controlled by inter- or intramolecular factors.

An interesting example of selective line-broadening in a liquid-phase spectrum is shown in Fig. 5.6. The radical was formed in aqueous solution by hydrogen atom abstraction (by ·OH) from 1,4-dioxane. As shown in Fig. 5.6, the two β-protons would be inequivalent in a static conformation, one proton giving a large overlap with the $2p$ orbital (axial) and the other a small overlap and hence a small hyperfine coupling (equatorial). Rapid inversion interchanges these protons, and in the fast inversion limit they would appear to be equivalent, giving a $1:2:1$ contribution. The single α-proton gives a constant, unmodulated doublet splitting. For intermediate inversion rates, the central doublet broadens prior to splitting out into four components, and in this spectrum they have become so broad that at first sight they might be overlooked completely.

Perhaps the most interesting linewidth effects are those that stem from chemical changes. A very large number of such effects have been studied, and several specific examples are given in the following chapters, especially Chapters 9 and 10.

5.6 Signs of hyperfine coupling constants from linewidths

Before leaving this section brief reference should be made to the way in which the signs of isotropic hyperfine coupling constants can sometimes be deduced from linewidth trends for radicals in the liquid phase. The method, which is widely used, really stems from the solid-state spectrum, and a knowledge of various aspects of the solid-state spectrum has to be assumed before any deductions can be drawn. First, it is clear that if the solid-state spectrum is obtainable, then the sign of A_{iso} can be fixed, provided it is assumed that the anisotropic tensor components take the signs $+2B -B -B$, as is generally the case for p-orbitals (but not for four of the five d-orbitals: Chapter 12). Instead of deriving equations to suit all situations, it is probably more instructive to illustrate the type of argument with a specific example. Consider the triplet spectrum ($I = 1$) shown in Fig. 5.7. In Fig. 5.7(a) this is shown for a radical in which $g_{\parallel} < g_{\perp}$ but having, hypothetically, an isotropic hyperfine coupling, which may be positive or negative. Now, for these two cases, consider the addition of an extra anisotropic coupling of form $|2B -B -B|$ such that the parallel ($2B$) coupling is coincident with $g_{\parallel}$. If A is positive, then the spectrum in Fig. 5.7(b)

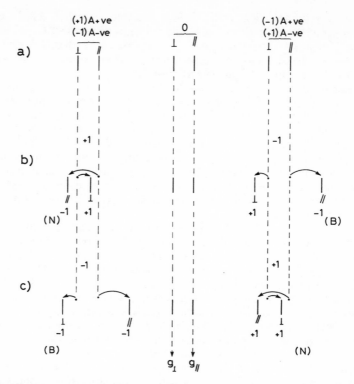

Figure 5.7 Method of estimating the sign of A_{iso} from the widths of isotropic lines. In (a) the solid-state features ($\parallel$ and $\perp$) are indicated for $I = 1$ with $g_{\parallel} < g_{\perp}$ $2B = 0$. The low-field feature is $M_I = +1$ if A is positive but -1 if A is negative. In (b) hyperfine anisotropy is introduced, with $2B$ positive, and A_{iso} positive. The low-field line is narrower than the high-field line. In (c) A_{iso} is negative, and the low-field line is broadened.

results, but if A is negative we get Fig. 5.7(c), for any magnitude of B. The point is that ΔB for the low-field line in (b) is less than ΔB for the high-field line, but this is reversed in (c). Thus in solution, if the radicals do not tumble too rapidly, the widths will be N→B for positive A and B→N for negative A. Thus provided one knows, or can reasonably guess, the form of the g-tensor components (i.e. $g_{\parallel} > g_{\perp}$ or $g_{\parallel} < g_{\perp}$) detection of a linewidth trend between $M_I = \pm n$ components leads to a prediction of sign. Generally, it is more straightforward to freeze the solution, but sometimes the concentration of radicals is too low to give a detectable solid-state spectrum.

This brief survey will, I hope, suffice for an understanding of the following chapters. For full treatments of linewidth and relaxation phenomena, see, for example, refs. 5.1 and 5.2.

CHAPTER 6

Examples from organic chemistry

Classification, like definition, is often arbitrary and one is wise to avoid it. Nevertheless, subdivision is essential, as is 'seeing the wood for the trees', so I will try to constrain this section to radicals that look 'organic', and those in Chapter 7 to those that look 'inorganic'. This leaves an overspill in Chapter 8 that I have called 'organo-inorganic' though 'organo-metallic' is a more usual term. Under 'organic' I place radicals containing C, H and first-row elements. The artificiality is illustrated by the appearance of $\cdot CF_3$ in this chapter and the isoelectronic $\cdot CO_3^{3-}$ in Chapter 7.

6.1 Alkyl radicals

We begin by returning to the methyl radical (Section 3.7). To what extent can we predict data for other alkyl radicals from the results for $\cdot CH_3$? Provided they all have a similar 'planar' σ-framework, the answer should be that there are close similarities. Comparison of the ^1H and ^{13}C isotropic coupling constants for the alkyl radicals in Scheme 6.1 show that this is indeed the case. The anisotropic hyperfine data for such radicals are less accurately known, but are certainly closely related. Since the $2B$ terms for ^{13}C are more directly informative for π-radicals, we can infer that the simple model, $R_3C\cdot$, which places unit spin in the $2p_z$ orbital on carbon, really is a good approximation.

Factors that contribute to the differences in A_{iso} (^1H and ^{13}C) for the alkyl radicals shown in Scheme 6.1 include:

   (i) Zero-point energy.
  (ii) Differences in the $2s$ and $2p$ content in the σ-orbitals and their distribution between the atoms.
 (iii) Possible deviations from planarity.
 (iv) Delocalisation onto the 'ligands'.

Perhaps because of the interesting structural implications, process (iii) has tended to dominate discussions of these differences, which are by no means fully understood. My own view, stemming from a wide range of data, is that (iii) is unimportant for the alkyl radicals listed in Scheme 6.1, and that these radicals can be treated as being effectively planar at the radical centre. Perhaps

SCHEME 6.1

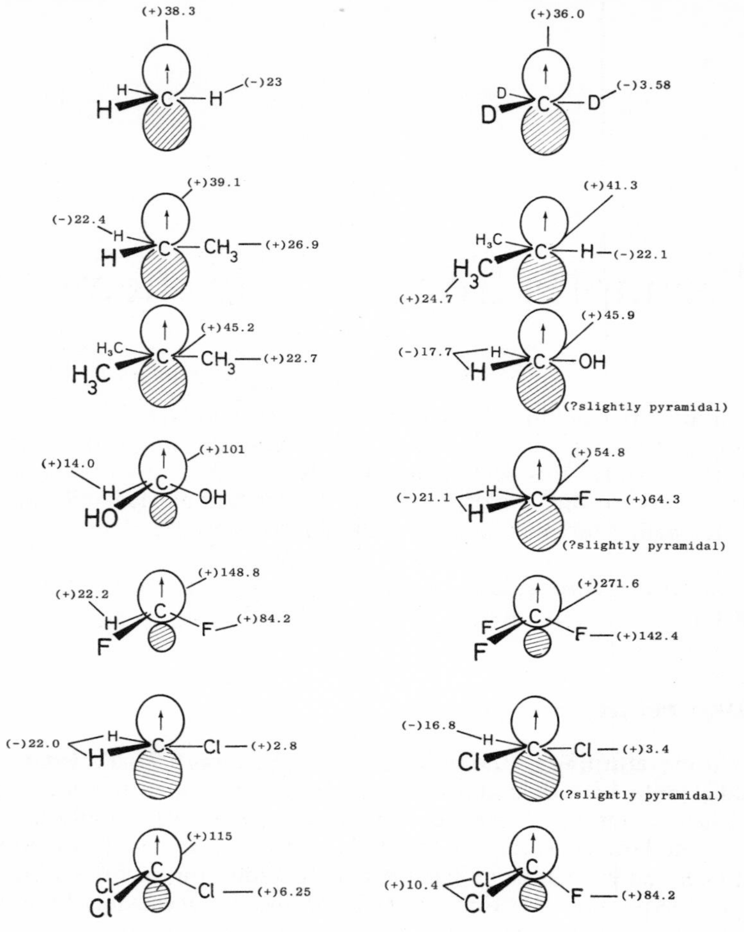

(+)38.3   H—C—H (−)23   H

(+)36.0   D—C—D (−)3.58   D

(−)22.4   (+)39.1   H—C—CH_3 (+)26.9   H

(+)41.3 H_3C—C—H (−)22.1 H_3C (+)24.7

(+)45.2 H_3C—C—CH_3 (+)22.7 H_3C

(+)45.9   (−)17.7 H—C—OH (?slightly pyramidal)

(+)14.0   (+)101   H—C—OH   HO

(+)54.8   (−)21.1 H—C—F (+)64.3 (?slightly pyramidal)

(+)22.2   (+)148.8   H—C—F (+)84.2   F

(+)271.6   F—C—F (+)142.4   F

(−)22.0 H—C—Cl (+)2.8

(−)16.8 Cl—C—Cl (+)3.4 (?slightly pyramidal)

(+)115   Cl—C—Cl (+)6.25   Cl

(+)10.4 Cl—C—F (+)84.2   Cl

the most compelling argument is that the pertinent e.s.r. data for $\cdot BH_3^-$ radicals and for $R_2\dot{C}OH$, $R_2\dot{C}F$ and $R_2\dot{C}Cl$ radicals are all remarkably similar to those for alkyl radicals. As discussed in Chapter 7, it seems that the difference between the central and ligand atom electronegativities is a major controlling factor governing the tendency for such radicals to become pyramidal. The radicals are clearly poised between the undisputably planar cations, R_3C^+ and the pyramidal anions, $R_3C:^-$ (isoelectronic with amines, $R_3N:$). There is no compelling reason why they should be planar, and many, such as $\cdot CF_3$ discussed below, are decidedly pyramidal, as is $\cdot CO_3^{2-}$. This is summarised in Fig. 6.1.

Good theoretical calculations have been forthcoming both supporting and disproving planarity for $Me_3C\cdot$ radicals. This serves to underline the subtlety of the problem, which still awaits definitive proof.

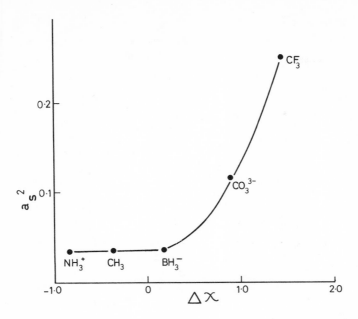

Figure 6.1 Trend in a_s^2 with the electronegativity difference $\Delta\chi$ for a variety of $\cdot AB_3$ radicals.

6.2 β-proton hyperfine coupling

The α-proton isotropic coupling for ethyl is slightly less than the $(-)23$ G expected for unit spin-density, and this can be taken to mean that slight delocalisation onto the methyl group has occurred. If this group is rotating rapidly with respect to the $-\dot{C}H_2$ group, the three protons will exhibit equal hyperfine coupling constants (fast averaging: see Chapter 5). This leads to an isotropic spectrum of $4 \times 3 = 12$ lines, as indicated in the stick diagram in Fig. 6.2(a). Note that the CH_3 protons (often called β-protons) have a coupling very close in magnitude to that for the $-\dot{C}H_2$ (α) protons. (This near equivalence lead to some confusion in the early years: solid-state spectra showed six features with nearly binomial intensity distribution, indicating five equivalent protons. A variety of formulations were put forward before it was realised that the 'classical' structure was satisfactory.) In fact, the coupling to the β-protons is positive whilst that to the α-protons is negative, but this sign difference cannot be gauged directly from the e.s.r. spectrum.

Although coupling to β-protons is often greater than that to α-protons, that to γ- and more distant protons is usually close to zero and rarely greater than 1 G. Thus, for example, the e.s.r. spectrum for the radical, $CH_3CH_2\dot{C}HOH$ shown in Fig. 6.2(b) is a doublet of triplets, with no sign of any splitting from the methyl protons.

Similar results have been obtained from $H\dot{C}(CH_3)_2$, $\dot{C}(CH_3)_3$ and many other alkyl radicals. The average β-proton coupling falls steadily to *ca.* 22 G on going to $\dot{C}(CH_3)_3$ probably because of slight delocalisation onto the methyl

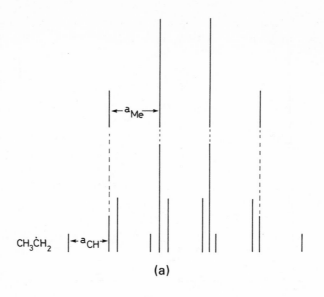

$\text{CH}_3\dot{\text{C}}\text{H}_2$

(a)

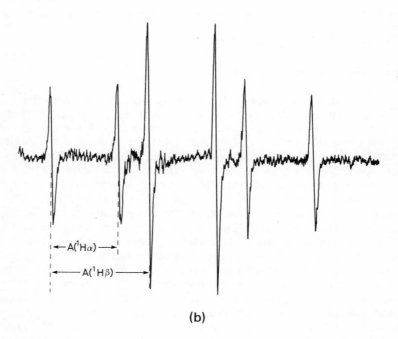

(b)

Figure 6.2 In (a) the way in which the CH_3 and CH_2 groups for the $\text{CH}_3\dot{\text{C}}\text{H}_2$ radical contribute to the total e.s.r. spectrum is indicated (note how, at low resolution, this appeared to be a simple sextet). In (b) a first-derivative e.s.r. spectrum for the $\text{CH}_3\text{CH}_2\dot{\text{C}}\text{HOH}$ radical is shown, the two sets of 1:2:1 triplets being well defined.

groups (Scheme 6.1). This delocalisation is thought to occur by a process called hyperconjugation or σ–π delocalisation. It is illustrated for the

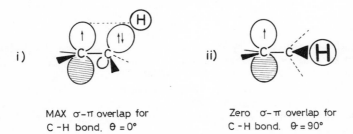

unit in Fig. 6.3 as a function of the relative orientation of the C—H bond. If this lies in the radical plane—that is, in the nodal plane of the $2p$ orbital—the coupling is expected to be close to zero, and to reach a maximum at 90° to this plane. This is expressed by the equation:

$$A(\beta H) = A \cos^2 \theta + B \qquad (6.1)$$

where $A \approx 46$ G and B is small if not zero.

This dependence of the hyperfine coupling of β-protons on angle has been amply verified. Generally, methyl groups rotate freely, giving a 1:3:3:1 quartet contribution, even at 77 K. However, in sterically crowded conditions this rotation may be quenched, in which case the three protons become inequivalent in a manner that appears to be arbitrary, depending presumably upon subtle crystal forces. Other groups, however, are more likely to exhibit restricted rotation or libratory motion only, at low temperatures. Thus for solid-state studies we must conclude that β-protons in alkyl radicals may exhibit hyperfine coupling constants anywhere between *ca.* 0 and 46 G, but that this will be nearly isotropic. (In fact the coupling is not strictly isotropic, but since the anisotropy falls off as $\langle r^{-3} \rangle$ it is small. Note that $A(\mathrm{Me})_\parallel < A(\mathrm{Me})_\perp$ for a rotating methyl group.)

Thus it seems that the qualitative theory of hyperconjugation adequately describes the hyperfine coupling of β-protons in alkyl radicals. More refined analyses suggest that positive spin-density is acquired by two mechanisms, both of which follow the $\cos^2 \theta$ law: one involves spin polarisation of the C—H σ-electrons and hence is not strictly a delocalisation, whilst the other is a real delocalisation. Hence it is not correct to use the measured coupling to β-protons to estimate the extent of delocalisation, the real delocalisation being probably closer to 50% of that estimated in this way [6.1].

If we accept that the methyl radical is planar then the proton hyperfine coupling takes on a special significance. This is because $\dot{\mathrm{C}}\mathrm{H}_3$ is then directly

i)   ii)

MAX σ–π overlap for   Zero σ–π overlap for
C–H bond. $\theta = 0°$ C–H bond. $\theta = 90°$

Figure 6.3 Overlap (σ-π or hyperconjugative) for a C—H bond in a radical $R_2\dot{\mathrm{C}}$—CHR_2' showing how this depends upon θ, the angle between this bond and the radical plane.

related to a whole family of π-radicals containing one or more planar, trigonal R_2CH groups. These include the $R_2\dot{C}H$ radicals having nearly unit spin-density on one carbon atom, discussed above, but also species such as $R_2C{=}CR_2{}^+$, $R_2C{-}CR{-}CR_2$ (allyl radicals) $C_6H_6{}^-$, etc. If we take $(-)$ 23 G as the ^1H coupling for unit spin-density, then the value obtained for a C—H proton in radicals such as these gives a fairly accurate measure of the effective spin-density on that carbon atom. This is because the electrons spend most of their time on, or near, the atoms involved in the conjugated systems, and spin on adjacent carbon atoms has little effect on the ^1H coupling concerned. (i.e. for a planar system, spin on $C_{(1)}$ and $C_{(3)}$ will not directly reach $H_{(2)}$ in the unit

This ceases to be true if the unit is non-planar.)

These factors form the basis of the very useful equation, applicable to such planar π-radicals:

$$A(H_i) = Q\rho_i \tag{6.2}$$

Here ρ_i gives the spin-density on the ith carbon to which H_i is bonded, and $Q \approx (-)$ 23 G. (This is usually termed 'McConnell's equation' [6.2], and it is a simple version of the more complex Equations 6.12 and 6.13 below.)

6.3 Allyl, vinyl and related radicals

A wide variety of π-radicals have been studied by e.s.r. spectroscopy, the simplest being allyl, $CH_2{-}CH{-}CH_2$. Simple expectation for this radical suggests a planar structure, with roughly sp^2 hybridisation at carbon, and with three electrons in the π-system. Two will be in π_1 having no extra nodal surface, and the unpaired electron will be in π_2, with a node through the central carbon atom. Thus we expect 50% spin density on the two outer carbon atoms, and using McConnell's equation (6.2) this predicts $A_{iso}(^1H)$ of $ca.$ $(-)11.5$ G for four equivalent protons. In fact, these four protons have $A_{iso} = -13.9$ G and there is a small coupling of $+4.0$ G to the C—H proton [6.3].

Thus the simple molecular orbital model is not far wrong. (Indeed, this is the theme of much of the data discussed herein. The way in which simple molecular orbital theory has been able to explain and even to predict the searching results obtained by e.s.r. spectroscopy has been a revelation. It is, indeed, an incredibly powerful theory.) The small deviations from this simple model are explained in terms of 'spin polarisation', largely of the two π_1 electrons. This puts negative spin-density (Chapters 3 and 4) on the CH carbon atom and hence an increased positive spin-density on $C_{(1)}$ and $C_{(3)}$. (This secondary effect is also well accommodated by relatively simple molecular orbital theory.)

Although the cation and anion of ethylene have not been studied, the cation $Me_2C{\doteq}CMe_2{}^+$ has been prepared [6.4]. The results ($A_{Me}{}^1H = 16.6$ G) illustrate a very important principle—that the extent of βC—H hyperconjugation increases markedly with positive charge. This suggests that for such radicals electron release from the C—H σ-electrons is important. The vinyl radical, $H_2C\!\!=\!\!\dot{C}H$, has been studied, and the results are well worth examining. There are two readily envisaged extreme structures ((a) and (b) in Scheme 6.2).

SCHEME 6.2

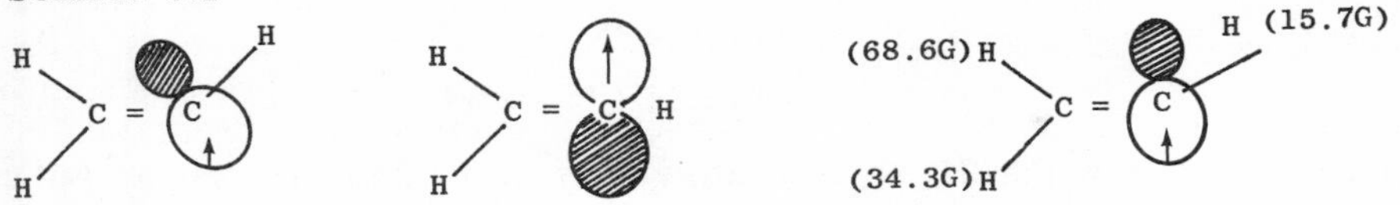

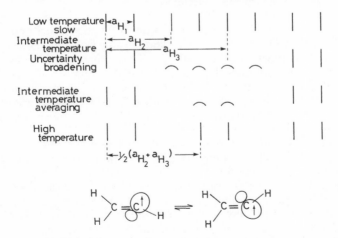

a)   b)   c)

The former (a) is the result of hydrogen atom removal with no subsequent change in shape; the unpaired electron is in an sp^2 hybridised σ-orbital (as in C_6H_5, discussed below). Alternatively, as in planar $\cdot CH_3$, the radical could move completely to the linear structure (b).

In fact, as is shown conclusively by e.s.r. spectroscopy, an intermediate structure (c) is adopted. When the α-proton is replaced by —CO_2H, the radical becomes 'linear', and the two β-protons give a coupling of 52 G, close to the average for those in $H_2C\!\!=\!\!\dot{C}H$. The 'linear' structure is, of course, found for $H_2C\!\!=\!\!\dot{C}^-$ which exhibits a proton coupling of 58 G for two equivalent protons [6.5]. The ^{13}C hyperfine coupling reflects these changes. Thus, for $H_2C\!\!=\!\!\dot{C}^-$, we find $A_\| = 77$ G and $A_\perp = 15$ G and $A_{iso} = 35.7$ G, typical of a $2p$ orbital on carbon, whilst A_{iso} for $H_2C\!\!=\!\!\dot{C}H$ is 107.6 G, showing an admixture of $ca.$ 10% $2s$-character as required by the 'bent' structure.

One interesting factor revealed by the e.s.r. results is that a rapid inversion occurs at the $\dot{C}H$ centre, and very low temperatures are needed to give the normal eight line spectrum. This broadening process is summarised in Fig. 6.4.

The anion, $H_2C\!\!=\!\!\dot{C}^-$, is isoelectronic with $H_2C\!\!=\!\!\dot{N}$. This, and its derivatives, $R_2C\!\!=\!\!\dot{N}$, are of wide occurrence as intermediates in the radical chemistry of

Figure 6.4 Stick representation for the e.s.r. features expected for $H_2C\!\!=\!\!\dot{C}H$ radicals at low and high temperatures. Note how, at intermediate temperatures, only the outer two doublets give narrow features.

organo-nitrogen compounds. The very large βH coupling of *ca.* 87 G makes $H_2C{=}\dot{N}$ and $RHC{=}\dot{N}$ radicals easy to detect by e.s.r. spectroscopy. The $H_2C{=}\dot{N}$ radical has also been detected in the radiolysis of aqueous cyanide ions, together with $H\dot{C}N^-$ [6.6]:

$$CN^- + e^- \rightarrow \cdot CN^{2-} \tag{6.3}$$

$$\cdot CN^{2-} + H_2O \rightleftharpoons H\dot{C}N^- + OH^-, \tag{6.4}$$

$$H\dot{C}N^- + H_2O \rightleftharpoons H_2CN\cdot + OH^- \tag{6.5}$$

These results show that $H\dot{C}N^-$ is a surprisingly strong base. The $H\dot{C}N^-$ radical exhibits an even larger ^1H coupling of *ca.* 125 G. It is isoelectronic with $H\dot{C}O$, the parent of another important class of radical intermediates, $R\dot{C}O$, and this also has a very large proton coupling of *ca.* 127 G.

These large isotropic proton coupling constants can be understood, for $H_2C{=}\dot{N}$, in terms of $\sigma{-}\pi$ overlap, enhanced by the short $C{=}N$ bond:

For $H\dot{C}O$ and $H\dot{C}N^-$ an additional factor arises because of the $2s$ contribution from carbon. This facilitates direct σ-delocalisation which puts positive spin-density on the hydrogen atom.

6.4 Some radicals containing nitrogen

This leads us to a brief consideration of other aliphatic radicals containing ^{14}N. These have been studied in the solid state using ionising radiation for their formation, and in most instances isotropic e.s.r. spectra have been obtained either from radicals generated in fluid solution, or using the adamantane matrix technique. The cation radicals $\cdot NR_3^+$ are directly analogous to alkyl radicals, the 'parent' $\cdot NH_3^+$ being isoelectronic with $\cdot CH_3$. In cases where one R group is hydrogen such cations are readily converted into amino radicals, $\cdot NR_2$, by proton loss.

Results for $\cdot NH_3^+$ are directly relatable to those for $^{13}\cdot CH_3$ (Scheme 6.1) and, as stressed elsewhere, help to establish that both species are 'planar' (apart from the unavoidable zero-point uncertainty). Comparison between $\cdot CMe_3$ and $\cdot NMe_3^+$ is interesting in that there is a large increase in hyperfine coupling to the β-protons on going from carbon (22.7 G) to nitrogen (27 G). This result sheds further light on the mechanism invoked for hyperconjugation, again suggesting that it is electron donation from the $C-H$ σ-orbitals that is dominant.

A radical whose structure you might not guess at first sight is the cation of triethylenediamine $(TED)^+$ (Fig. 6.5) (Scheme 6.3).

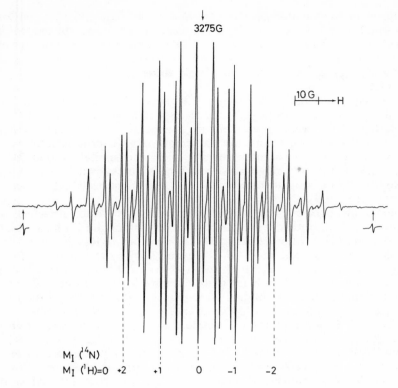

3275G

10 G → H

M_I (^{14}N)

M_I (^{1}H)=0   +2   +1   0   −1   −2

Figure 6.5 First-derivative X-band e.s.r. spectrum for the radical cation of tetraethylenediamine, showing a set of quintets from the two equivalent nitrogen nuclei.

The results for TED$^+$ are unique in that they show that the two nitrogen atoms and all twelve protons are magnetically equivalent [6.7]. A liquid-phase spectrum is shown in Fig. 6.5. Solid-state studies show clearly that the nitrogen atoms remain equivalent at 77 K, which almost certainly means that the minimum energy structure is one in which the spin is equally distributed rather than being centred on one nitrogen and undergoing a rapid charge transfer to

SCHEME 6.3

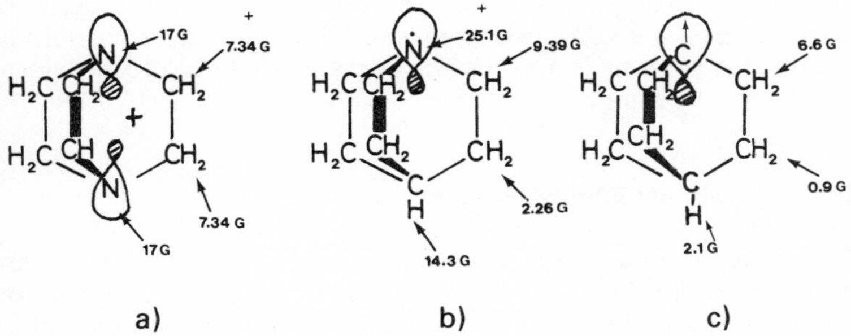

a)   17 G   7.34 G H_2C CH_2 CH_2 H_2C CH CH_2 7.34 G   17 G

b)   25.1 G   9.39 G H_2C CH_2 CH_2 H_2C CH_2 CH_2 H   2.26 G   14.3 G

c)   6.6 G H_2C CH_2 CH_2 H_2C CH_2 CH_2 0.9 G   H   2.1 G

49

give apparent equivalence 'on the e.s.r. time-scale'. The ^{14}N isotropic coupling is too small for a structure in which the electron is confined to two sp^3 hybrid orbitals on the nitrogen atoms. Delocalisation onto the CH_2 groups undoubtedly accounts for some spin-density, but the results suggest some flattening at the two nitrogen atoms which would bring them together slightly and increase the overlap.

The bridge-head radical shown in 6.3(b) has also been studied. The ^{14}N isotropic coupling of 25.1 G [6.8], is, as expected, larger than that for $\dot{N}R_3^+$ radicals (ca. 19 G) because the structure cannot acquire complete planarity at the nitrogen atom. Also note the extremely large δ-proton coupling which presumably again reflects direct overlap. For the similar carbon radical in Scheme 6.3(c) the ^{13}C bridge-head coupling is not known, but the δ-proton coupling of 2.7 G is again unusually large and suggests direct delocalisation across the ring [6.9].

Amino radicals, $\cdot NR_2$, are isoelectronic with $\cdot CR_2^-$ radicals which do not seem to have been studied by e.s.r. spectroscopy although they have often been invoked as reaction intermediates. On going from $\cdot NR_3^+$ to $\cdot NR_2$ radicals, there is a significant fall in $A_{iso}(^{14}N)$, even though the unpaired electron is largely confined to nitrogen in both (ca. 19 G → ca. 14 G). One facile explanation is that spin polarisation is far more effective for the strongly attenuated σ electrons than for the non-bonding 'lone-pair' in $\cdot NR_2$. Indeed, if the lone-pair were to be neglected one would expect a reduction of ca. $\frac{1}{3}$, which is close to that observed. However, another, related, contribution may be centred in the hybridisation of the orbitals at nitrogen since the bond angle for $\cdot NR_2$ is less than the 120° required for the $\cdot NR_3^+$ species. This may account, for example, for the real increase in $A_{iso}(^{14}N)$ on going from $\cdot NH_2$ to $\cdot NR_2$ (ca. 12 G → ca. 14 G). If the $R\hat{N}R$ angle is greater than that of 103° for $\cdot NH_2$, the $2s$ contribution should be greater and hence also the isotropic coupling.

For $R_2\dot{C}$—NR_3^+ radicals, ^{14}N acquires negative spin-density primarily via spin-polarisation of the $\dot{C}$—N σ-electrons. The isotropic coupling of ca. $(-)4.0$ G reflects this. On going to $R_2\dot{C}$—$\dot{N}R_2$ species the possibility of π-delocalisation arises and this should modify $A_{iso}(^{14}N)$. In fact, the isotropic coupling remains small (ca. 5 G) suggesting that the classical structure, $R_2\dot{C}$—$\ddot{N}R_2$ is a reasonable approximation of the true structure [6.10]. However, the α-proton coupling is reduced to about 15 G suggesting that up to 30% delocalisation may occur. Probably the configuration about ^{14}N remains pyramidal. This result can be compared with that for the $H_2\dot{C}$—BMe_2 species obtained from BMe_3 by γ-radiolysis. Here again delocalisation is to be expected, giving $(\pi_1)^1$ rather than $(\pi_1)^2(\pi_2)^1$, and a planar radical is now expected. Nevertheless, the coupling of ± 4.0 G to ^{11}B suggests that the classical structure is again a good approximation, and the hyperfine coupling for the methylene protons is now 18 G, suggesting only ca. 15% delocalisation [6.11].

6.5 Some radicals containing oxygen

We now turn to a consideration of some oxygen radicals. $R_3\dot{O}^{2+}$ and even $R_2\dot{O}^+$ species have not been characterised by e.s.r. spectroscopy, although the latter should be preparable and readily detectable. Simple alkoxy radicals such

50

as $H_3C\dot{O}$ have not yet been reported, but certain complex molecules of biological significance have been converted into RO· radicals and their e.s.r. spectra are discussed in Section 13.3. Further consideration of these elusive species can be found in a discussion of the parent ·OH radical in Section 7.4.

However, the radicals $R_2\dot{C}OH$, $R_2\dot{C}OR$, $R_2\dot{C}O^-$ and $R_2\dot{C}OH_2^+$ are well known, as, of course, are the related stable nitroxide radicals, $R_2\dot{N}O$. (A typical $R_2\dot{C}OH$ spectrum is shown in Fig. 6.2(b).) These radicals are probably still nearly planar at the carbon centre (Scheme 6.1). However, introduction of another oxygen is enough to give a pyramidal radical. Thus electron addition to RCO_2^- or RCO_2H gives $R\dot{C}O_2^{2-}$ and $R\dot{C}O(OH)^-$ radicals whose isotropic ^{13}C coupling of ca. 100 G is certainly too large for a planar radical and implies a direct admixture of $2s$ character only possible for the pyramidal species. These radicals have been prepared by electron addition (via ionising radiation) to carboxylic acids or their salts [6.12]. Similar radicals, $R\dot{C}(OH)_2$ or $R\dot{C}(OR)_2$, have been prepared by selective hydrogen atom abstraction using, for example, ·OH or ·OR radicals [6.13] (Scheme 6.1).

6.6 Nitroxide and iminoxy radicals

The ketyl radicals, $R_2\dot{C}O^-$, show a strong tendency to dimerise, and of course the conjugate acids, $R_2\dot{C}OH$ dimerise irreversibly. In contrast, $R_2\dot{N}O$ radicals dimerise far less readily and when the R groups are bulky they form stable liquids or solids (cf. the dimerisation of CO_2^- and NO_2, Section 7.4). Because of their stability, the nitroxides, especially di-t-butyl nitroxide, $(Me_3C)_2\dot{N}O$, have been very widely studied, not so much per se, but as environmental probes. Aspects of their use are outlined in Chapters 9 and 13. As with $R_2\dot{C}OH$ radicals, they are nearly planar, but X-ray crystallographic studies suggest a slightly pyramidal structure in some cases.

Another remarkably stable class of radical is termed 'iminoxy', with the general structure $R_2C{=}\dot{N}O$. The unpaired electron is formally in an in-plane 'σ' orbital on nitrogen, and because this is strongly $2s$–$2p$ hybridised, $A_{iso}(^{14}N)$ is large (ca. 31 G) [6.14]. There is some tendency to dimerise on cooling, the resulting dimer probably having a weak N—N bond as in N_2O_4. Indeed, these radicals are in many ways comparable with NO_2, discussed in Section 7.4.

6.7 Spin-trapping

Despite the evident abundance of radicals directly detected and studied by the e.s.r. technique, there are many circumstances in which direct detection of radical intermediates is still impossible. This may arise because the stationary concentration is too low or because the radicals give broad, undetectable features as, for example, the alkoxy radicals, RO· (Chapter 13). The 'spin-trapping' technique relies upon a radical addition to some substrate to give a new, stable radical, whose concentration can therefore accumulate to the level required for detection.

Such secondary radicals are usually nitroxides because of their stability, and because the lines are usually so narrow that small splittings from protons present in the added radical can be detected.

51

A single example must suffice: consider the radiolysis of methanol. This, on general chemical expectation, should proceed by the ionisation:

$$CH_3OH \rightarrow CH_3\dot{O}H^+ + e^- \qquad (6.6)$$

followed by the rapid proton transfer

$$CH_3\dot{O}H^+ + CH_3OH \rightarrow CH_3\dot{O} + CH_3OH_2{}^+ \qquad (6.7)$$

and H-atom transfer,

$$CH_3O\cdot + CH_3OH \rightarrow CH_3OH + H_2\dot{C}OH \qquad (6.8)$$

Thus $CH_3\dot{O}$ radicals seem to be necessary intermediates, but solid-state e.s.r. spectra only reveal the presence of $H_2\dot{C}OH$ radicals. However, in the presence of a nitroxide precursor, RNO, several different radicals appear, one being clearly $HOCH_2$—N(O)—R and the other RN(O)OMe. The former, a normal nitroxide, exhibits a $1:2:1$ triplet from the methylene protons, and the latter, being a derivative of a nitro-radical, gives a far larger coupling to ^{14}N since such radicals are pyramidal at nitrogen. Thus, both $H_2\dot{C}OH$ and $\dot{O}CH_3$ are proven intermediates.

6.8 Some radicals containing fluorine

Radicals containing α-fluorine atoms are of interest because of the large doublet splitting from ^{19}F. $R_2\dot{C}F$ radicals are probably planar or nearly so (cf. Scheme 6.1), but the large anisotropy shown by $A(^{19}F)$ ($2B_z = 200$ G) shows that there is appreciable π-delocalisation onto fluorine despite the large electronegativity difference. Further substitution by fluorine leads to marked bending, as shown by $A_{iso}(^{13}C)$. The increase in $A_{iso}(^{19}F)$ is also characteristic of bending. The proton coupling, negative for H_2CF, is surely positive for $H\dot{C}F_2$.

Interaction with a β-fluorine atom, as in $H_2\dot{C}CH_2F$, for example, follows the same $\cos^2 \theta$ law discussed for β-hydrogen atoms. Thus it seems to be controlled by $\sigma-\pi$ overlap, but, relative to C—H or C—C hyperconjugation, the C—F bond is generally disfavoured. This means that the preferred conformation for $H_2\dot{C}CH_2F$ is one in which the fluorine tends towards the radical plane:

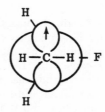

This is clearly demonstrated by the e.s.r. results which show two β-proton coupling constants that are greater than the average and that increase on cooling. From this temperature dependence the energy barrier to the relative rotation of the —CH_2 and —CH_2F groups can be estimated.

6.9 Aromatic radicals

Although most molecules can neither gain, nor lose, σ-electrons with any facility, they can often gain and lose π-electrons with considerable facility, and this particularly applies to aromatic molecules, the larger the ring system, the greater the ease of both gain and loss of electrons. (There is a tendency for the electron-loss species to form dimers, such as $(C_6H_6)_2^+$, but the anions do not exhibit this relatively weak interaction. Thus these π-systems follow the general rule that a shortage of electrons leads to sharing.) The results for the ions of molecules such as naphthalene or anthracene are straightforward and simple molecular orbital theories for π-systems reproduce them very well. A typical spectrum, that for the naphthalene anion, is shown in Fig. 6.6. For the benzene ions, however, the high symmetry leads to an interesting case of degeneracy, similar to those which are frequently encountered with transition metal complexes.

The simple form of the π molecular orbitals expected for a completely symmetrical benzene anion are shown in Scheme 6.4(a) and (b). Although not obviously so, the (a) and (b) molecular orbitals are exactly equal energetically,

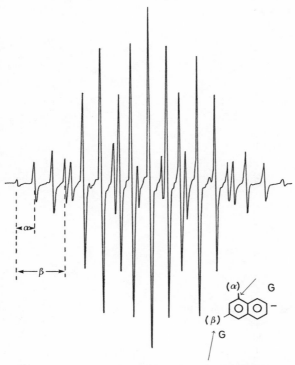

Figure 6.6 First-derivative X-band e.s.r. spectrum for the naphthalene anion, showing five sets of quintets from the α- and β- protons.

53

but become different if the symmetry is slightly reduced by a distortion such as an elongation. Such a distortion would therefore be predicted by the Jahn–Teller theorem, which shows that there is always a small net stabilisation when the degeneracy of two levels is lifted in this manner. However, at normal temperatures, both cations and anions have spectra indicative of six equally coupled protons (and, from ^{13}C features, the carbon atoms are also equivalent). Thus there must be a rapid equilibrium between the distorted forms so that a 'dynamic Jahn–Teller' distortion is occurring. Evidence that this is indeed occurring has been adduced from linewidth measurements.

Introduction of substituents in the ring can, of course, lift the degeneracy permanently. For example, consider the cation and anion of durene (Scheme 6.4(c) and (d)). The e.s.r. spectra show unambiguously that the structure in which the four equal methyl groups overlap with the orbital of the unpaired electron to a maximum is favoured for the cation, and to a minimum for the anion. These are indicated in Scheme 6.4 and are understandable in terms of the electron releasing power of the methyl groups. This fits in with the enhanced β-proton hyperfine coupling for the cation (11.0 G) compared with the coupling expected relative to that for the ethyl radical (i.e. $26.9 \div 4 = 6.72$ G). This is another example of the increase in hyperconjugation with cationic charge at the radical centre (cf. $Me_2C\!\!=\!\!CMe_2^+$ discussed above).

Scheme 6.4

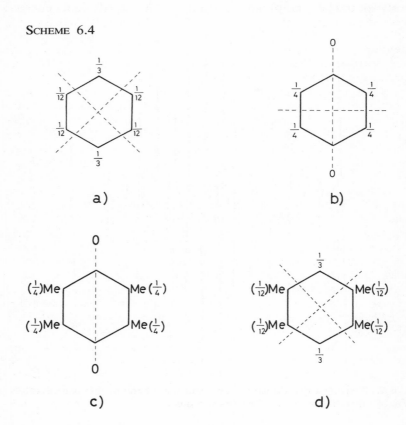

a)

b)

c)

d)

54

The simple equation (6.2) for these aromatic systems is satisfactory for proton hyperfine coupling. However, for ^{13}C coupling or ^{14}N coupling to ring nitrogen nuclei, spin-density on the two adjacent ring atoms plays an important rôle in controlling the total isotropic hyperfine coupling. This is discussed in Section 3.9 where it is shown how helpful Q- or U-values can be and how necessary it is to consider the adjacent atom terms.

Many other interesting but unexceptional aromatic anions and cations have been studied. In general, substituents behave in a predictable manner. The NO_2 group lends stability, and $PhNO_2^-$ and the meta and para dinitrobenzene anions have been extensively used in solvation studies (see Chapter 9). Since the aliphatic nitro-anions, RNO_2^-, are definitely pyramidal at nitrogen, there is a real chance that $ArNO_2^-$ anions will also be pyramidal, although to a less extent. Solid-state studies give the principle values of the ^{14}N hyperfine coupling, and from these the $2p:2s$ ratio turns out to be *ca.* 15, which is only slightly less than that expected for a planar system [6.16].

Another interesting experiment is to have two alkyl groups ortho to the NO_2 group, thus forcing it to twist out of the plane of the benzene ring. In the limit of a 90° twist the excess electron would need to select a site between the ring and the NO_2 group. In fact, there is a fall in spin-density in the ring and a gain at nitrogen, showing that, in the limit, the nitro-group would be chosen.

The semiquinones, especially *p*-benzosemiquinone and its derivatives are also relatively stable, and are of particular importance in biology. They too have been used extensively for solvation studies (Chapter 9). The e.s.r. results, especially the ^{17}O and ^{13}C coupling constants for the carbonyl groups, show that the major spin-density is located thereon, and that environmental changes alter the distribution between carbon and oxygen, leaving the spin-density in the ring largely unchanged (see Section 9.10).

It is generally true to say that modern molecular orbital theories reproduce the experimental results for these planar aromatic π-radicals quite well. However, from time to time the unexpected does crop up, and one example of this will be mentioned here. It is, of course, the unexpected that makes work of this sort exciting. Monofluorobenzene and certain difluorobenzenes add electrons to give quite normal π-anions, as is illustrated by the isotropic data given in Scheme 6.5. However, hexafluorobenzene gives a species containing the expected six equivalent fluorine atoms, but exhibiting a far larger hyperfine coupling to fluorine than one would ever predict [6.17]. The problem is exacerbated by the fact that the apparent anisotropy is remarkably small. Thus the $2s$ character is far greater than expected for a normal spin-polarisation phenomenon, and the π spin-density as measured by the $2B$ term is far smaller. At the time of writing, these results have not been satisfactorily explained. One can talk about σ- rather than π-delocalisation, or about some major distortion from the planarity of the ring of the parent molecules [6.18], but why should there be such a drastic switch to a new and unexpected structure? Maybe the answers are known as you read this; if not, then why not speculate?

Two other important radicals derived from aromatic systems are the cyclohexadienyl and phenyl radicals (Scheme 6.6, (a) and (b)). The former (a), prepared by H-atom addition to benzene, is remarkably stable, thus establishing that such structures represent potential energy minima rather than being transition states as has been supposed in the past. The e.s.r. spectra show two

SCHEME 6.5

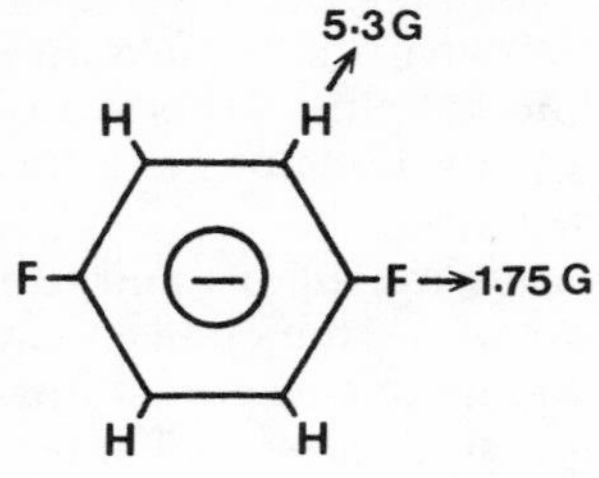

(1)

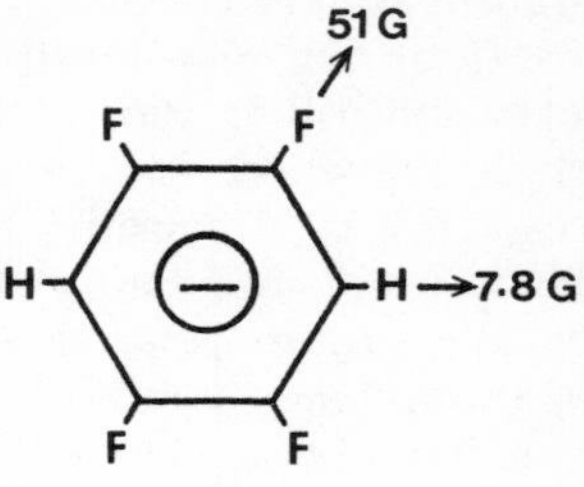

(2)

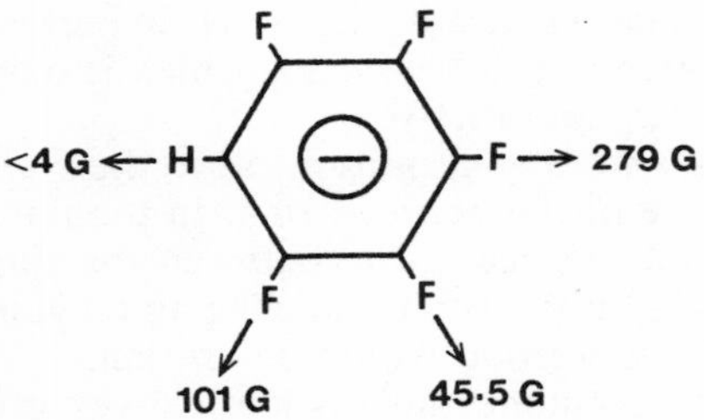

(3)

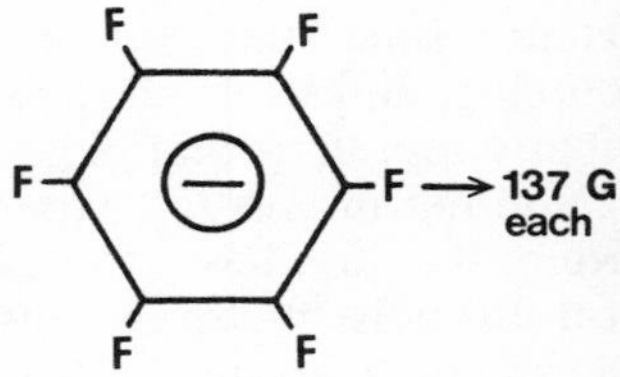

(4)

Note that for (1) the anion is 'normal' and there seems to be a node through the two fluorine substituents.

very strongly coupled protons from the $>$CH$_2$ group (*ca.* 50 G) three more weakly coupled protons (*ca.* 10 G) from the ortho and para-positions, and two very weakly coupled protons (*ca.* 3 G) from the meta positions. The ring-coupling constants are typical of suitably substituted aromatic radicals, but the very large coupling to the CH$_2$ protons is characteristic of radicals in this class, and is clearly greater than one would expect, using the spin-densities on the two adjacent (ortho) carbon atoms and the $\cos^2 \theta$ law. The explanation given by Whiffen [6.19] is a quantum mechanical reinforcement from the presence of two interacting centres of high spin-density. It is an extremely useful phenomenon since it gives rise to a well defined 'fingerprint' spectrum for such radicals, which might otherwise be difficult to detect.

SCHEME 6.6

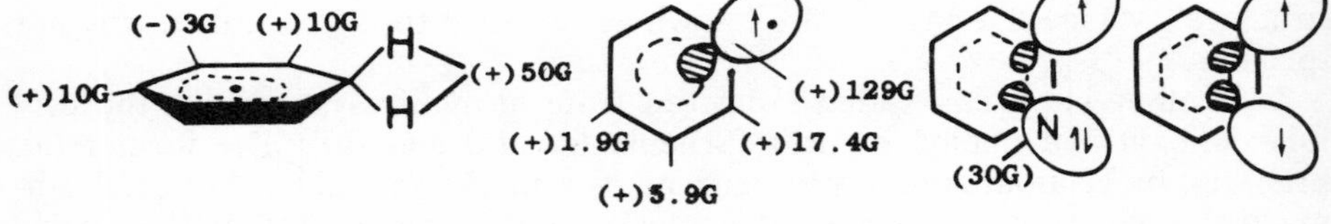

a)

b)

c)

d)

Phenyl radicals (Scheme 6.6(b)) do not have this advantage, their spectra being confined to a smaller field-range and hence more likely to be obscured by features from other radicals. Also, they are extremely reactive species. Nevertheless, they have been prepared, for example, in rare-gas matrixes, by reactions such as:

$$PhCl + e^- \rightarrow Ph\cdot + Cl^- \tag{6.9}$$

and the characteristic e.s.r. spectrum shows two strongly coupled protons (18 G) and two weakly coupled protons (6.4 G). The remote para proton coupling is only 1.9 G, suggesting that direct overlap across the ring is not very important. The isotropic ^{13}C coupling for this radical should be large because in this case the ring structure prevents any major distortion towards linearity at the radical centre, and this is indeed the case, the coupling of 129 G corresponding to *ca.* 0.12 spin-density. (An sp^2 orbital with no delocalisation corresponds to 0.33.)

One interesting example of a hetero-aromatic radical is the pyridyl radical (6.6(c)) [6.20]. The spectrum for this radical shows that although the major spin-density remains on carbon, nevertheless, there is considerable delocalisation onto nitrogen, by what has been termed pseudo π-overlap, presumably similar to that which occurs in the benzyne molecule (6.6(d)).

This brief survey only begins to touch on the many interesting organic radicals that have been studied by e.s.r. spectroscopy. However, many others of interest to the organic chemist are discussed in Chapters 8, 9 and 10. For further reading, see the excellent reviews in the Chemical Society's Special Publications, ESR Vols. I, II, III and IV, and also reviews by Gilbert and Norman [6.21] and by Fischer [6.22].

CHAPTER 7

Examples of inorganic radicals

Of the few stable radicals familiar to all chemists, perhaps NO, NO_2, NF_2 and ClO_2 would top the list. This chapter is concerned with the information that e.s.r. spectroscopy can provide for such species. We begin with atoms and work through diatomic, triatomic and tetraatomic to pentaatomic species, bringing in more complex radicals at appropriate places. Before setting out on this journey, which is described far more thoroughly in a previous book [7.1], we take a brief look at those interesting entities, solvated and trapped electrons.

7.1 Solvated and trapped electrons

When sodium interacts with liquid ammonia, a homogeneous blue solution is formed which in the dilute region has all the properties expected for a $1:1$ electrolyte solution; the sodium atoms giving Na^+_{solv} and 'solvated electrons', e^-_{solv}. The species responsible for the blue colour, $(\nu_{max} \approx 6700\,\text{cm}^{-1})$ the solvated electron, can be thought of as an electron occupying a cavity in the solvent of the same size and structure as that which normally surrounds a monatomic ion such as I^-.

This model closely resembles that of the F-centre in alkali halide crystals, shown in Fig. 7.1. Most solvents react too rapidly with the alkali metals to give stable solutions of this type, but certain amines and ethers do, and in particular hexamethylphosphoramide, HMPA, $(Me_2N)_3PO$, gives relatively stable solutions. However, using the technique of pulse radiolysis, transient spectra for solvated electrons can be obtained from many other solvents, including water. Also, radiolysis of certain glassy solids, including aqueous salt solutions, alcohols, amines and ethers results in the formation of violet or blue colours associated with trapped electrons. These are again similar to F-centres and are probably trapped at vacancies or imperfections in the glass.

e.s.r. spectroscopy confirms the presence of a paramagnetic species, but for the fluid solutions the single narrow feature obtained is uniformative since all the expected hyperfine features from the nuclei of solvent molecules defining the cavities are washed out by extremely rapid solvent exchange. The glassy solids are better, but only in the sense that deductions can be drawn from linewidths which are now quite large. However, a wealth of detail has been

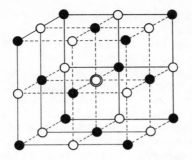

Figure 7.1 Structure of an F-centre in an alkali halide crystal. ○ Anions; ● Cations; ◎ Electron.

gleaned from F-centres in alkali halide crystals, and these confirm the model shown in Fig. 7.1 most convincingly. Thus hyperfine coupling to six equivalent cations accounts for some 50% of the spin (in the outer s-orbitals of these ions) and there is clear hyperfine coupling to the first shell of anions. Indeed, using ENDOR spectroscopy, details of several more remote shells has been obtained. These very small interactions cannot be used directly to map the outer regions of the wave-function of the electron however, because a part, if not all of the coupling must stem from polarisation effects. Thus the results show that the electrons are strongly confined to the cavity itself and to the outer s-orbitals of the six adjacent cations.

Since the solvent molecules under discussion have no low-lying acceptor orbitals such as those for the alkali metal cations, it seems that electrons in such media should meet with greater resistance to penetration outside the trapping cavity, and e.s.r. results seem to confirm this. Thus, for example, in concentrated glassy solutions of salts in water little or no width increment is found from the cations or anions despite their inevitable close proximity. In contrast, the water protons make a major width contribution, which arises because of an orientation polarisation of the cavity molecules. This is nicely proven by an experiment using alcohol solvents. Irradiation at 4.2 K gave a narrow e.s.r. singlet suggesting little width enhancement by the OH protons. On warming to 77 K the width increased to the normal value indicative of strong OH proton coupling [7.2]. This increase was retained on recooling. This surely means that no orientational polarisation exists initially, but the negative electron charge is capable of breaking hydrogen bonds and re-orienting the solvent molecules rapidly at 77 K.

7.2 Atoms

Although most molecular radicals cannot be detected by e.s.r. spectroscopy when in the gas phase, most paramagnetic atoms are readily detectable, and chemists often use this technique as an accurate method of monitoring their concentration. In the liquid or solid states, interesting matrix or solvent effects are observable which have been extensively studied. Minor matrix effects can be explained reasonably well with current theories and are not our present concern. Major changes have been detected when atoms are generated within

59

the matrix or solvent rather than being 'matrix isolated' from the gas phase. One example must suffice.

Isolation of silver atoms from the gas phase in a water matrix at 77 K gave e.s.r. parameters close to the gas phase. However, formation of such 'atoms' by irradiating frozen aqueous glasses containing Ag^+ ions gave several slightly different centres having hyperfine coupling constants reduced by *ca.* 20%, and g-values in the 1.999 region rather than that of the free-spin [7.3]. This is interpreted in terms of relatively strong bonding to several water molecules causing delocalisation and a consequent admixture of $5p$ orbitals into the $5s$ level on silver. One interesting facet is that solvation is not 'switched off' when the atom is formed in the glass, and the other is that a range Ag^0 centres implies a range of Ag^+ hydrates rather than a single fixed structure in solution.

7.3 Diatomic radicals

Of the very wide range of diatomics studied by e.s.r., I have selected the following to illustrate the information given by e.s.r. studies:

$$CN, NO, N_2{}^-, F_2{}^- \text{ and } Ag_2{}^+$$

All these can be discussed in terms of the simple bonding scheme in Fig. 7.2. The radical CN, formed for example from HCN, has its unpaired electron in

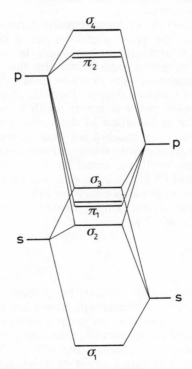

Figure 7.2 Energy level scheme for diatomic radicals.

σ_3, which, because of its antibonding character, has its major density on carbon. This is confirmed by the e.s.r. results which show that the spin is almost completely confined to an $s-p$ hybrid orbital on carbon. Indeed, the isotropic ^{14}N coupling is very small and probably negative [7.4], showing that spin polarisation wins over direct delocalisation in σ_3. This is a surprising result which taken alone seems to imply that σ-delocalisation is far less important than π-delocalisation when the nuclei have different electronegativities. On going to NO, which has two more electrons ($\ldots \sigma_3^2 \pi_2^1$), the e.s.r. results show that the unpaired electron spends *ca.* 70% of its time on N and the rest on oxygen. Again, because the orbital is antibonding, the electron is forced onto the less electronegative atom. (Or one can reverse the argument and say that the result establishes the antibonding character of the orbital.)

Of course, for N_2^-, which is formed from the azide ion by radiolysis, the electron is evenly distributed in the π_2 orbital, and it is interesting to consider the significance of the hyperfine coupling constants. (The g-tensor for NO and N_2^- resembles that for O_2^+, discussed in Chapter 2. The form is again characteristic of the π_2^1 configuration.) The $2B$ value is only 10.6 G, suggesting some 37% spin-density on each nitrogen. This low result could mean that the $2B^0$ value (33 G) is somewhat large, but it is unlikely to be in error to the extent. I expect that it is low because the radical is not stationary, but librating about the z axis. This will partially average out the anisotropy for the same reason that free rotation averages it to zero. For small light molecules such motion is the rule rather than the exception, and one has to beware of the effect, since it always reduces $2B$ below the true value. Symmetrical motion of this type leaves the A and g-tensors axial, but an asymmetric motion will introduce a spurious asymmetry that can again be dangerously misleading.

The isotropic coupling of +1.4 G for ^{14}N in N_2^- is again very low. For example, NH_3^+ has $A_{iso} = 19.5$ G, from which one might predict 9.7 G for N_2^-. A major reason for the small value is the 'back-bond' spin-polarisation effect (see Chapter 3). If we argue that π-spin on N_1 polarises the σ electrons to give, say, $+\alpha$ in the N_1 2s-orbital, then it will give *ca.* $-\alpha$ in the N_2 2s-orbital; similarly for spin on N_2. Thus one might guess that the net value would be zero, which is very nearly true. In fact, the experimental value is close to half that for nitrogen atoms ($3.7/2 = 1.85$ G) which means that inner shell polarisation may be the major contributor in this case.

On going to F_2^- we return to a σ radical, this time with one electron in σ_4, which to a good approximation is simply $[2p_z(1)-2p_z(2)]$. This radical is one of many $hal_{(1)}-hal_{(2)}^-$ radicals which are labelled V_K centres by physicists. They are the primary hole-centres in irradiated alkali halide crystals, and are formed from the halogen atoms in the manner shown in Fig. 7.3. The e.s.r. results leave absolutely no doubt as to their identity, structure and mode of formation. Briefly, for F_2^-, A_{max} lies along the three F—F directions, as does the free-spin g-value. The $2B$ value corresponds closely to 50% spin-density, and A_{iso} is small.

The radical Ag_2^+ resembles F_2^- in being a σ^* radical, but brings us back to the bottom of the energy level diagram in Fig. 7.2, since now we have, simply, a single electron in σ^1. These cations are readily formed when silver atoms are formed in the presence of Ag^+ ions:

$$Ag^0 + Ag^+ \rightarrow Ag \overset{.}{-} Ag^+ \tag{7.1}$$

61

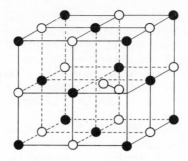

Figure 7.3 Structure of the F_2^- radical in an alkali-metal fluoride. ○ F^-; ● Cations; ○—○ F_2^-.

The e.s.r. results show clearly that the two silver atoms are equivalent and that the electron is essentially in the two $5s$ orbitals, the $5p$-character being small.

Before turning to triatomic radicals, I should mention the ·OH radical, since this is of extreme importance in much radiation, photochemical and redox chemistry. In liquid-phase studies this radical is never detected, even though it is clearly involved. Also, in much solid-state work it is not detected. However, in ice, and in many rigid aqueous solutions, ·OH radicals do have well defined e.s.r. spectra. These are π^* radicals, resembling O_2^-, but in this case it is the hydrogen-bond framework in the ice crystals that lifts the degeneracy of the $2p$ orbitals, the bonds being formed preferentially to the p orbital that is filled, thus isolating the half-filled orbital. The proton hyperfine coupling is quite normal for a π-radical. See also Chapter 13 for a discussion of RO· radicals.

7.4 Triatomic radicals (AB₂)

The dioxy radicals, OAO, illustrate the sort of arguments that are usually invoked when e.s.r. is used in the task of identification. Let us consider the effect of ionising radiation on a nitrite salt. Two major magnetic centres are formed, both showing a triplet indicative of hyperfine coupling to ^{14}N, and chemical expectation is that NO_2 and NO_2^{2-} are responsible. Ionising radiation is used extensively in the preparation of radicals in solids. The dominant initial step is electron loss (Chapter 10). If the cations are small enough, this stage might be represented as a hole in the valence band together with an electron in the conduction band, and one would expect a relaxation back to the ground-state. In fact, however, it seems that localised states readily occur, the electron becoming trapped at a specific anion after becoming thermalised. Also the hole remains on one anion as NO_2. This is primarily because both NO_2 and NO_2^{2-} have different dimensions from those of NO_2^-, so that electron transfer is severely limited by the need to match shapes before transfer, which represents a considerable barrier.

So our task is to decide which e.s.r. spectrum should be assigned to a particular radical. This can readily be accomplished by reference to elementary expectation for the two radicals. Consider first the 16-electron cation, ONO^+. This is linear, the lowest empty orbital being the two-fold degenerate π_3^* $(\bar{\pi}_u)$ orbital shown in Fig. 7.4(a). The addition of one electron causes the molecule

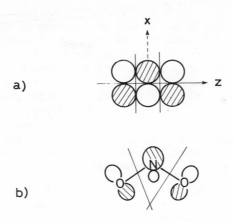

x

z

a)

b)

Figure 7.4 (a) The $\pi_3^*(x)$ orbital for NO_2^+. (b) The related pseudo π-orbital for NO_2.

to bend ($\theta = 134°$ for NO_2), and this splits the π orbitals into two, one ($\pi_y : 2\bar{b}_1$) remaining pure π in character, the other ($\pi_x : 3a_1$) taking on the bent character of the molecule by adding $2s$-character on nitrogen as depicted in Fig. 7.4(b). This admixture stabilises the orbital so that the added electron is to be found therein.

A second electron will pair with this, giving NO_2^-, but a third will go into the pure π orbital ($2\bar{b}_1$).

Thus, one radical should exhibit a large isotropic coupling to ^{14}N and the other a very small coupling arising via spin polarisation. Just this is observed: The radical that is clearly NO_2 has $A_{iso} = 54.7$ G and $2B = 13.2$ G whilst the other (NO_2^{2-}) has $A_{iso} = 13.0$ G and $2B = 19.3$ G. Of course, in this particular instance, NO_2 is a stable species and can be prepared chemically. This has been done, and the identification confirmed. However, the general procedure for identification is clear, and such simple structural arguments often lead to unambiguous assignments. Having decided on these, one can come back and use e.s.r. results to map out the orbital in some detail. So for NO_2, we deduce *ca.* 50% spin-density on ^{14}N, with a $2p/2s$ ratio of *ca.* 4. For NO_2^{2-} the orbital is pure $2p_y$ on nitrogen, the population being again *ca.* 0.5. These conclusions have been confirmed and extended by the study of ^{17}O hyperfine coupling from radicals enriched in this isotope, which is in too low abundance (0.037%) to be detected without enrichment. Although delocalisation onto oxygen in NO_2 can in principle occur via 'π' type or σ-orbitals, the former appears to be favoured.

It is interesting to compare results for members of an isoelectronic series. For CO_2^- and NO_2 the ^{13}C and ^{14}N coupling constants show that the unpaired electron is more confined to the central atom for CO_2^- than for NO_2. This shows that an antibonding orbital is indeed involved. The extensive delocalisation onto oxygen for NO_2 is one of the reasons why the equilibrium:

$$N_2O_4 \rightleftharpoons 2NO_2 \tag{7.2}$$

is important at room temperature, whereas that involving the oxalate anion:

$$C_2O_4^{2-} \rightleftharpoons 2CO_2^- \tag{7.3}$$

is not.

63

The directional nature of the orbitals involved should also be stressed. e.s.r. spectra for single crystals show that for CO_2^-, for example, $A_\parallel$ falls along the H—C direction of the parent formate ions. For NO_2^{2-}, the parallel direction is perpendicular to the planes of the parent nitrite ions. These results agree with the structures, so everything fits.

These directions are reflected in the structures of the dimers. Thus N_2O_4 has the planar structure (7.4) to be expected for optimum overlap but when the 19-electron SO_2^- dimerises a completely different structure (7.5) results, which again reveals the distribution of the unpaired electron in the monomer. For such 19-electron radicals, dimerisation is once more encouraged by localisation on the central atom. Thus $S_2O_4^{2-}$ dissociates only to a minor extent, but Cl_2O_4 is unknown.

$$
\begin{array}{cc}
\text{(7.4)} & \text{(7.5)}
\end{array}
$$

Nineteen-electron AB_2 radicals studied by e.s.r. spectroscopy are especially common, including the relatively stable NF_2 and O_3^- centres, together, for example, with N_3^{2-}, $FClO^-$ and $ClOCl^+$. Twenty-one-electron species such as F_3 or ClO_2^{2-} do not seem to have been studied. However, surprisingly, the 23-electron species, F_3^{2-}, is relatively well known, being a significant radiation product in certain alkali fluorides. Being one electron short of three fluoride ions, the bonding must be weak, and the bond presumably stays formed because the crystal structure favours the movement of two neighbouring F^- ions to stabilise a fluorine atom rather that the normal pairwise movement to give F_2^-. The e.s.r. results again accord well with expectation for this species, the ^{19}F hyperfine coupling revealing the σ^* nature of the unpaired electron's orbital and the concentration of spin on the central atom [7.1].

7.5 Tetraatomic radicals (AB_3)

It is interesting to compare the results for the set of six hydrides, BH_3^-, CH_3, NH_3^+, and AlH_3^-, SiH_3, PH_3^+. The results for the former trio show that these radicals are either all planar, or just possibly very slightly pyramidal. Thus the A_{iso} values for ^{11}B, ^{13}C and ^{14}N all correspond to ca. 3.5% $2s$-spin-density. This is to be expected if spin polarisation of the σ-electrons, together with the zero-point energy effect for planar radicals, is all that is involved. Otherwise, if we were to take the A_{iso} values as indicative of non-planarity, the deviation must be tiny in all three radicals (see Chapter 6) [7.5].

One of the most compelling explanations for bond angle trends is based upon electronegativity differences. For the families under consideration this suggests that the angle of deviation from planarity should increase as the central atom electronegativity falls. This trend, clearly seen for the AlH_3^-, SiH_3, PH_3^+ set (Scheme 7.1), occurs because of the drift of the σ-bonding electrons away from the central atom as the electronegativity falls. This drift, for the planar radical, depletes the $2p_{x,y}$ and the $2s$ orbitals of electrons. However, for the 90° radical it depletes the $2p_{x,y,z}$ orbitals and not the $2s$

SCHEME 7.1

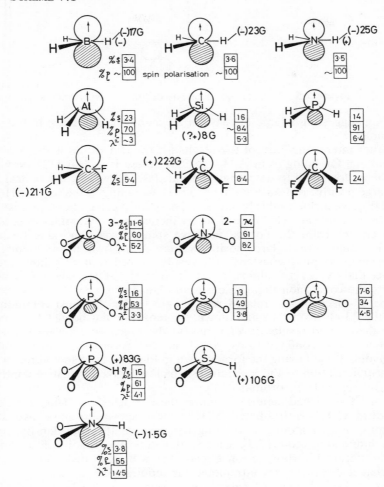

orbital. Since the atom will strongly resist loss from the deeper $2s$ level, there is a steady tendency to reduce the $2s$ character in the σ-orbitals and hence to increase it in the unpaired electron's orbital as the electronegativity of the central atom falls (Scheme 7.1).

We conclude that the BH_3^- set all fit on a region in which bending has not begun to be significant, whilst the AlH_3^- set are in a region in which bending is extensive. Thus the simple electronegativity argument does not seem to carry through from the first to the second row, not surprisingly.

A similar picture arises for the 25-electron radicals CO_3^{3-}, NO_3^{2-}, and PO_3^{2-}, SO_3^-, ClO_3. Now both groups are pyramidal because of the relatively high electronegativity of oxygen, but the extent of bending is far less for CO_3^{3-} and NO_3^{2-} than for the PO_3^{2-} set (Scheme 7.1). Changing the ligand gives concordant results: on going from CO_3^{3-} to CF_3 there is a large increase in

65

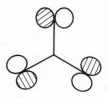

Figure 7.5 The non-bonding $1a_2'$ orbital for $NO_3 \cdot$ radicals.

$A_{iso}(^{13}C)$, suggesting a large increase in $2s$ character and hence in the pyramidal character.

Another interesting facet of the results for the trihydrides is the trend in the ^1H coupling. On going from AlH_3^- to PH_3^+ there is a clear fall, the value for PH_3^+ being too small to detect in the solid state. However, the radical $H\dot{P}O_2^-$ has a large positive proton coupling (+87 G). Coupling in the series CH_3, CH_2F and CHF_2 give a clue to this behaviour. Theoretical calculations suggest that on going from a planar AH_3 radical to an increasingly pyramidal one, $A_{iso}(^1H)$ should change from the normal negative spin-polarisation value (usually *ca.* -20 G) through zero, to large positive values. Apart from the ambiguity in sign for the small coupling constants, the results accord well with this prediction. The fact that $A_{iso}(^1H)$ is almost constant for BH_3^-, CH_3 and NH_3^+ therefore supports the conclusion that these are all planar.

These 25-electron and related radicals are by far the most common, but a few in the 23-electron class are known, the most familiar being CO_3^- and NO_3. Again, theory and results dovetail nicely: the unpaired electron is predicted, and found, to be confined to oxygen. If no distortion is present the orbital is non-bonding ($1a_2'$) having the form shown in Fig. 7.5. Spin should reach A only via spin-polarisation of the σ-electrons which will give a negative contribution in the $2s$, $2p_x$ and $2p_y$ orbitals [7.1].

Use of ^{17}O-enriched materials confirm these conclusions. Thus, for example, the radical CO_3^- in irradiated $KHCO_3$ was shown to have two strongly interacting oxygen atoms, accounting for most of the spin-density, and one weakly interacting atom [7.7]. Clearly the radical is readily distortable, but since the parent bicarbonate ion already has a built-in distortion one cannot argue that it is a necessary intramolecular requirement.

A few radicals having 27 valence electrons have also been studied. Thus $\cdot ClO_3^{2-}$ and $\cdot BrO_3^{2-}$ have their unpaired electrons quite strongly localised on the central atom, and have the low symmetry expected, given the presence of a sterically active 'lone-pair' of electrons. This is more clearly illustrated by results for the isoelectronic SF_3 molecule, which has two equivalent fluorine atoms with $A_{iso}(^{19}F) = 54.3$ G and one having $A_{iso} = 40.4$ G [7.8]. No accurate view of the geometry of these radicals is yet available.

7.6 Pentaatomic radicals (AB₄)

There are two major classes of interest to e.s.r. spectroscopists, 31-electron radicals typified by PO_4^{2-}, having the unpaired electron confined to a non-bonding orbital on the ligands, and 33-electron radicals, such as PO_4^{4-} or PH_4, in which the unpaired electron has a high density on the central atom.

66

The former have e.s.r. properties closely resembling those of the tetraatomic analogues such as NO_3. Thus the electron, formally delocalised onto all four ligands in an undistorted molecule (T_d) becomes largely confined to three, two or even one ligand p-orbital depending on the distortion experienced by the radical. The spin then interacts with the nucleus of the central atom via spin polarisation of the σ-bonding electrons to give a small negative isotropic coupling that is a function of the s-character in these orbitals. The extent of anisotropy obviously depends upon the degree of confinement, being maximum for spin on one ligand and zero for complete delocalisation. Deviations from the free-spin g-value again depend upon the extent of delocalisation, and hence the g-tensor is not a characteristic of a given radical. For example, PO_4^{2-} radicals in different crystalline environments exhibit a large variety of g-tensor components, and even $A_{iso}(^{31}P)$ is quite variable although always small.

Turning to 33-electron AB_4 radicals, the radical PH_4 has had a chequered history, but has recently been unambiguously identified by its liquid-phase e.s.r. spectrum (Fig. 7.6) [7.9]. This radical is the simplest of a class of radicals known to organic chemists as phosphoranyl radicals. The first to be studied was PF_4, but a wide range of such radicals has recently been studied both in the liquid and solid phases, and the results are considered in some depth in Section 8.2.1, since they illuminate a number of important principles.

The isotropic ^{31}P hyperfine coupling is very sensitive to the nature of the ligands, with $\cdot PH_4$ and $\cdot PF_4$ being at opposite ends of the range, with $A_{iso}(^{31}P) = 519\,G$ for PH_4 and $ca.$ $1300\,G$ for PF_4. Clearly, the electronegativity of the ligands plays an important rôle. This, at least in part, reflects the antibonding nature of the orbital involved. The more electronegative the ligands the more the antibonding electron is confined to the central atom.

In both cases, there is clear evidence for a distortion from T_d towards a 'trigonal-bipyramidal' structure. Thus for $\cdot PH_4$ two hydrogen atoms couple remarkably strongly with $A_{iso}(^1H) = 199\,G$ and the other two hardly at all ($A_{iso} = (\pm)\ 6.0\,G$). Theory for this molecule suggests that the two strongly

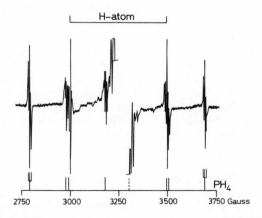

Figure 7.6 First derivative X-band e.s.r. spectrum assigned to $\cdot PH_4$ radicals (obtained from γ-irradiated PH_3 in neopentane at 100 K by A. J. Colussi, J. R. Morton and K. F. Preston, *J. Chem. Phys.*, 1975, **62,** 2004).

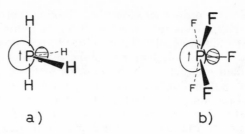

Figure 7.7 Predicted structures for (a) $\cdot PH_4$ and (b) $\cdot PF_5^-$ (or SF_5).

coupled atoms are nearly axial, with the other two equatorial, as shown in Fig. 7.7. The same structure is invoked for $\cdot PF_4$, and a nearly linear arrangement for the axial ligands and the phosphorus atom has been revealed by a study of this radical in a single crystal of PF_3 [7.10]. Delocalisation is again onto these axial ligands, but is now largely via the $2p_\sigma$ orbitals and hence $A_{iso}(^{19}F)$ is relatively small.

It is interesting to compare the results for $\cdot PH_4$ discussed above, with those for a radical obtained from the photolysis of PH_3, also thought originally to be $\cdot PH_4$ [7.11]. Although the spectrum appeared to comprise a very large doublet from ^{31}P hyperfine coupling (974 G), there were central features that, in our view [7.12], belong to the outer 'doublet' features, the whole spectrum constituting a 'triplet' which we assign to $P_2H_6^+$ formed by loss of an electron. It is known [Section 8.4] that $R_3P\cdot^+$ radicals react rapidly with R_3P molecules to give $R_3P\dot{-}PR_3^+\sigma^*$ radicals, and hence the same is expected for $H_3P\cdot^+$ cations. The parameters, including the new g-value, agree well with expectation for $P_2H_6^+$.

7.7 AB₅ and AB₆ radicals

Exposure of hexafluorophosphate salts to ionising radiation gives $\cdot PF_5^-$ [7.13]:

$$PF_6^- + e^- \rightarrow \cdot PF_5^- + F^- \qquad (7.4)$$

but for many years this centre was thought to be $\cdot PF_4$, since the e.s.r. spectrum only revealed the presence of four fluorines. The fact that these were equivalent gave a clue, since for genuine $\cdot PF_4$ they are never found to be equivalent, but this was explained away in terms of a rapid pseudo-rotation. However, careful examination of the spectrum has now revealed a small extra doublet splitting from the fifth ligand. As with $\cdot PF_4$ the coupling to ^{31}P is large (*ca.* 1350 G). Similar results for SF_5 led to an initial identification as SF_4^+. The expected structure of these species is shown in Fig. 7.7(b).

Irradiated SF_6 gives a species containing six equivalent fluorine ligands in addition to $\cdot SF_5$, and this must surely be the parent anion, SF_6^- [7.14]. This anion exhibits an extremely large $A_{iso}(^{33}S)$ of 643 G, equivalent to a $3s$ population of *ca.* 0.66. It is not yet known if the ligands are genuinely equivalent or if there is an equilibration that is fast on the e.s.r. time-scale. The very high $3s$-character on sulphur suggests the former, and solid-state studies would be of great interest for this species.

7.8 Cyclic phosphazenes

Before leaving this section on inorganic radicals, I should mention another result that seems to have considerable structural significance. Cyclic phosphazenes such as $N_3P_3Cl_6$:

are generally supposed to have some 'aromatic' character stemming from π-type ring delocalisation of the $6p(\pi)$ electrons formally on nitrogen, via suitable $3d$-orbitals on phosphorus. If this π-structure were of major significance, electron gain or electron loss might be expected to have similarities with these reactions for benzene and other aromatic molecules.

Our e.s.r. results [7.15] show conclusively that electron gain results in the formation of a normal phosphoranyl radical, the excess electron being confined to one phosphorus atom and to two (axial) chlorine atoms. There was no evidence for any ring-delocalisation. Similarly, but less definitively, the electron loss centres appeared to have the unpaired electron confined to a single nitrogen atom. These results do not disprove the π-delocalisation hypothesis, but taken together with results for $R_2\dot{C}—\overset{+}{P}L_3$ radicals (Section 8.1) they make extensive delocalisation appear to be very improbable.

CHAPTER 8

Some organo-inorganic radicals

In this chapter, we meet a range of radicals having much in common with those mentioned in Chapters 6 and 7. The title 'organo-metallic' is a more common, but, to my mind, less suitable title for many of these species. The inorganic elements involved are taken from the second and subsequent rows of the periodic table, going backwards from the halogens. We consider two major types of radical, one mainly carbon-centred, and the other with the unpaired electron primarily centred on the inorganic group.

8.1 Carbon-centred α-radicals

The group of radicals $R_2\dot{C}Cl$, $R_2\dot{C}Br$ and $R_2\dot{C}I$ have proven to be remarkably elusive to e.s.r. spectroscopists. This is primarily because the quadrupole moments of the halogen nuclei are relatively large, and there are strong, asymmetric electric fields at these nuclei. These electric fields exert an aligning force on quadrupolar nuclei that may oppose the magnetic aligning forces (Fig. 8.1). This obviously upsets the clear-cut situation that usually holds in the absence of such coulombic forces, and indeed can render the resulting e.s.r. spectra almost unrecognisable. In the liquid phase the effect is to broaden the hyperfine features, but the normal pattern of hyperfine lines remains. However, in the solid state, the normal pattern results only when the electric and magnetic fields are aligned, or when the magnetic interaction dominates.

The way in which the spectrum is calculated to change for magnetic fields along the principle axes for a nucleus with $I = \frac{3}{2}$ (chlorine or bromine) is shown in Fig. 8.2(a) for an $R_2\dot{C}Br$ radical. For further details, see, for example, ref. 8.1. When the effect of the electric field dominates, the spectra should have the appearance of a triplet rather than a quartet and this has been confirmed for α-chloro and α-bromo radicals along the y axis. The powder spectrum for a typical α-bromo radical is shown in Fig. 8.2(b): complete interpretation is clearly difficult, but the x-features are well defined. Despite the difficulties involved in a full interpretation, such powder spectra are of great use as 'fingerprints' in radiation studies.

The results for these radicals, once extracted, are unexceptional. Delocalisation onto halogen is found to be relatively small, and hence spin polarisation of

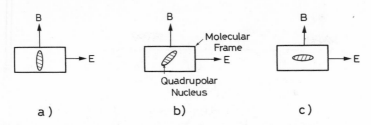

Figure 8.1 The joint effect of an internal electric-field gradient (E) and an external applied magnetic field (B) on a quadrupolar nucleus. (a) E or Q is small and the nucleus lines up with the magnetic field. (c) E and Q are large and the nucleus lines up with the electric field. (b) is an intermediate case.

the $\dot{\text{C}}$—hal σ-electrons is significant. This gives a negative contribution to A_{iso}. This explains why A_{iso} values are close to zero. A typical result for $H_2\dot{\text{C}}Cl$ radicals is as follows:

$$A_{iso} = +2.8\,\text{G}, \qquad 2B \approx 17.7\,\text{G}$$

Hence the spin-density on chlorine is approximately $17.7/100 = 0.177$.

Radicals containing α-sulphur or selenium are more difficult to identify because of the low abundance of magnetic isotopes. ^{33}S is only 0.74% abundant which means that a given feature for the ^{33}S species is only about 0.0018 as intense as that for the ^{32}S species, even if it has the same width. Since enriching in ^{33}S is an expensive experiment, most studies have relied upon the measurement of the g-values and ^{1}H hyperfine coupling for identification. The results are reasonable: the small spin-density in the $3p_x$ sulphur orbital confers a g-shift for g_z and to a less extent for g_y, making $g_{av} \approx 2.0049$ [8.2]. The α- and β-proton coupling constants are typical for alkyl radicals with a minor degree of delocalisation.

Phosphorus or arsenic atoms α- to a carbon radical centre are potentially more interesting structurally, since we can have either $R_2\dot{\text{C}}PL_2$ or $R_2\dot{\text{C}}PL_3$ type centres. For the former, direct π-delocalisation is possible, but not for the latter, unless $d(\pi)$–$p(\pi)$ bonding is invoked. (There will, of course, be a small hyperconjugative interaction with the P—L σ-orbitals.) In fact, just such $d(\pi)$–$p(\pi)$ bonding is extensively postulated in other areas of chemistry, and for molecules such as R_2CPR_3, the structure is usually depicted as $R_2C\!=\!PR_3$. If this were a reasonable representation, loss of an electron to give $(R_2CPR_3)^+$ radicals would leave an unpaired electron in the π-orbital. In fact, many such radicals have now been studied by e.s.r. spectroscopy, and all have the properties expected for the representation $R_2\dot{\text{C}}PR_3{}^+$. Thus the coupling to αH or βH in the $R_2\dot{\text{C}}$ moiety is quite normal, as are the g-tensor components. The coupling to ^{31}P is small and negative, and can be adequately explained in terms of a remote dipolar coupling and spin polarisation of the C—P σ-electrons (see, for example, ref. 8.3). There is no need to invoke $d(\pi)$–$p(\pi)$ bonding, so any such interaction must be of minor importance. I conclude that the zwitterionic representation, $R_2\bar{\text{C}}$—$\overset{+}{\text{P}}R_3$ is more satisfactory for the parent molecules, but considerable charge neutralisation via σ-electron polarisation is to be expected.

71

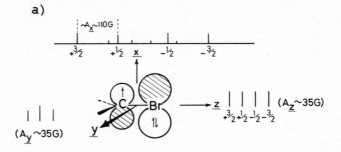

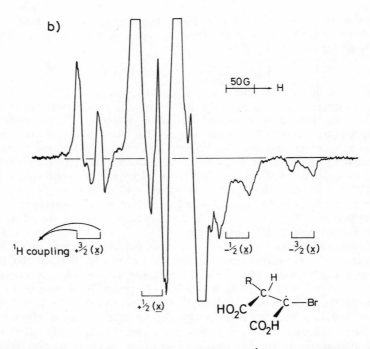

Figure 8.2 (a) Form of the e.s.r. spectrum for an $R_2\dot{C}Br$ radical showing predicted hyperfine features from ^{81}Br (or ^{79}Br) together with data obtained from single crystal studies. (b) The X-band powder spectrum for

radicals in irradiated bromomaleic acid, showing the well defined and characteristic quartet of doublets giving $A_x(Br)$.

72

The results of e.s.r. studies show that even for $R_2\dot{C}PR_2$ radicals, delocalisation is relatively unimportant. This is established by the small ^{31}P coupling and the small reduction in ^1H coupling in the alkyl groups. For fourth-row elements the same picture emerges. For example, $R_2\dot{C}SiR_3$, $R_2\dot{C}GeR$, $R_2\dot{C}SnR_3$ and $R_2\dot{C}PbR_3$ radicals have all been studied, and their e.s.r. spectra fit these formulations very well [8.4]. As with the $R_2\dot{C}{-}PR_3^+$ radicals, there is no case for invoking any significant $d(\pi)-p(\pi)$ bonding. In my view, these results constitute one of the most searching pieces of experimental evidence against the general concept of major stabilisation by $d(\pi)-p(\pi)$ bonding in such molecules.

8.2 Carbon-centred β-radicals

These are perhaps even more interesting in the bonding information that they provide than the α-radicals. This is because there is a general tendency for such radicals to favour the conformation shown in Fig. 8.3. This conformational preference may not be very large, and solution spectra often exhibit a temperature dependence indicative of a restricted oscillation about the $\dot{C}{-}C$ bond (indicated in Fig. 8.3(b)). Cooling leads to an increase in A_{iso} for the

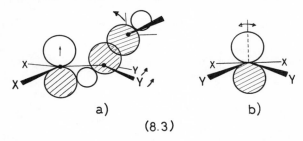

a)   b)

(8.3)

Figure 8.3 Preferred conformation for β-groups (X) that tend to give $\sigma-\pi$ conjugation (hyperconjugation) in radicals $R_2\dot{C}{-}C(R_2)X$.

β-heavy atom substituent and a fall in A_{iso} for β-protons. The upper limit for $^1H(\beta)$ is, of course, that for free rotation (ca. 20–25 G), whilst the lower limit is expected to be that for the rigid conformation shown in Fig. 8.3, which according to the $\cos^2 \theta$ law (Section 6.2) is ca. $50 \times \cos^2 (60)$, i.e. ca. 12.5 G. In fact, $A_{iso}(^1H_\beta)$ is sometimes considerably less than this, even when $A_{iso}(^1H_\alpha)$ indicates that delocalisation is not very extensive. This may arise because of a distortion of the 'tetrahedral' arrangement about the β-carbon, as indicated in Fig. 8.3(a).

Some results are summarised in Fig. 8.4, taken from both liquid- and solid-state studies. Results for the halogens are only clear in the case of β-chlorine. These show that the symmetrically bridged structure (8.5), which

$$\underset{\text{(8.5)}}{>\dot{C}\!\cdots\!\dot{C}<}$$

Cl

73

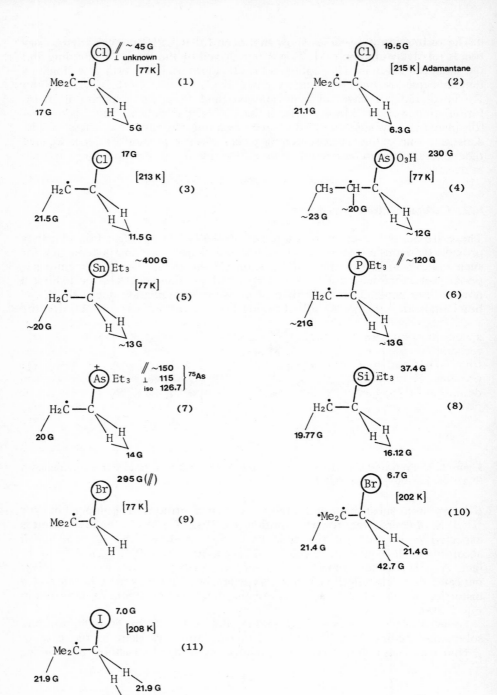

Figure 8.4 Scheme showing a range of β-substituted alkyl radicals which exhibit σ–π delocalisation. Some e.s.r. data are indicated.

has been frequently invoked, is incorrect, and that the structures in Figs. 8.3 and 8.4 are reasonable [8.5]. However, our results for the radical

$$(CH_3)_2\dot{C}\overset{\displaystyle Cl}{\overset{\displaystyle /}{-}}CH_2$$

[8.6] have established that chlorine migration is extremely rapid, even at 77 K. This is because this radical was unambiguously formed from γ-irradiated t-butyl chloride, the only reasonable mechanism for this being:

$$(CH_3)_3C\!-\!Cl \xrightarrow{\;\gamma\;} H_2\dot{C}C(CH_3)_2Cl \qquad (8.1)$$

$$H_2\dot{C}\!-\!C(CH_3)_2Cl \xrightarrow{\;fast\;} H_2C\overset{\displaystyle Cl}{\overset{\displaystyle |}{-}}\dot{C}(CH_3)_2 \qquad (8.2)$$

The results given in Fig. 8.4(9) for $Me_2\dot{C}\overset{\displaystyle Br}{\overset{\displaystyle /}{-}}CH_2$ radicals, formed from $Me_3C\!-\!Br$ by γ-radiation [8.7] in addition to being poorly defined, are not universally accepted, and it may be of interest to explain why. Firstly, the case for their formation hinges entirely on the detection of the very broad e.s.r. spectrum shown in Fig. 8.5. We felt that these anisotropic features were reasonably assigned to hyperfine interaction with ^{79}Br and ^{81}Br (each giving four lines), and the resolution on the $M_I = +\frac{1}{2}$ line seemed to establish the presence of two methyl groups; nevertheless, compared with the beauty and definition of most liquid-phase e.s.r. spectra, this certainly leaves a lot to be desired. This is well illustrated by comparison with the spectrum shown in Fig. 8.6. This spectrum was obtained from isobutylbromide (Me_2CHCH_2Br) present in dilute solution in solid perdeutero adamantane, after exposure to X-rays [8.8]. It shows well defined hyperfine coupling to ^{79}Br and ^{81}Br, together with coupling to several ^1H nuclei. The interpretation of the proton coupling given

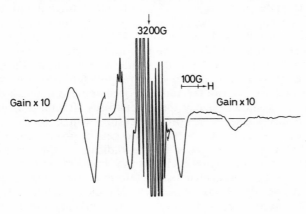

Figure 8.5 First derivative X-band e.s.r. powder spectrum assigned to $Me_2\dot{C}\!-\!CH_2Br$ radicals formed in γ-irradiated $Me_3C\!-\!Br$ at 77 K.

75

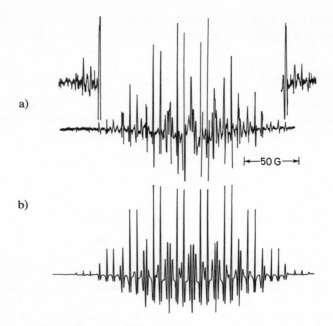

a)

b)

|←—50 G—→|

Figure 8.6 Second derivative e.s.r. spectrum for a radical obtained by X-irradiation of 1-bromo-2-methylpropane in adamantane at 202 K (a) and a computer-simulated spectrum (b). This is assigned to the radical $(CH_3)_2\dot{C}CH_2Br$ by R. V. Lloyd and D. E. Wood. (Taken, with permission, from *J. Amer. Chem. Soc.*, 1975, **97**, 5986.)

in ref. 8.8 is indicated in Fig. 8.4(10), the spectrum being assigned to the β-bromo radical $Me_2\dot{C}CH_2Br$. In order to explain the very small hyperfine coupling to bromine $[A(^{81}Br) = 6.7\ G]$, the structure in Fig. 8.3 was dismissed, the bromine atom being thought to lie quite close to, but not exactly in, the radical plane. Perhaps because of the clear definition of this spectrum, this assignment has met with wide acceptance. However, the analysis requires the curious coincidence that $A(^1H)$ for the two methyl groups and one of the CH_2 protons is exactly half $A(^1H)$ for the other CH_2 proton. Furthermore, this fortuitous equality is maintained over a range of temperatures. Even more curious is the fact that a similar radical, obtained from the iodine analogue (Fig. 8.4(11)) which shows clearly the six features stemming from coupling to ^{127}I $(I = \frac{5}{2})$, also shows a set of ten proton features which, on this analysis, again requires a factor of exactly two between two sets of proton coupling constants. This means that the bromine and iodine radicals adopt precisely the same asymmetric conformation. My own view is that bromine and iodine are unlikely to differ from chlorine, and are almost certain to adopt the structure in Fig. 8.3. (There is a considerable body of indirect chemical evidence for such structures.) I think that the ten hyperfine features indicate the presence of nine equivalent protons, and that possible candidates for the species present in adamantane are $(CH_3)_3C\cdot \ldots Br^-$ and $(CH_3)_3C\cdot \ldots I^-$. The proton coupling constant of 21.4 G is satisfactory for $(CH_3)_3C\cdot$ radicals, and weak interaction with halide ions have been observed previously [8.9] (see Section 10.2.4), but this does require that $(CH_3)_3C\cdot$ radicals be formed. In fact, uncomplexed $(CH_3)_3C\cdot$ radicals were detected on annealing [8.8], which lends credence to the suggestion. This unresolved controversy is included here because of its intrinsic interest, and because I wish to avoid giving the impression that everything is clear-cut in this field.

There are a few liquid-phase results that suggest that β sulphur groups also tend to favour the conformation shown in Figs. 8.3 and 8.4. However, for phosphorus and arsenic, the results are unambiguous, both from liquid- and solid-state studies [8.10,8.11]. The liquid-phase results, which serve to finger-print the species, show that $A_{iso}(^{31}P)$ is relatively large (Table 8.1), showing direct admixture of the $3s$ (phosphorus) orbital, and the β-proton coupling is small. Solid-state spectra were generally broad and the anisotropy poorly defined except for the radical

$$CH_3\dot{C}HCH_2\text{---}As\overset{\displaystyle O}{\underset{\displaystyle OH}{\overset{\displaystyle O}{\diagdown}}}$$

The results show that admixture of an orbital of *ca.* sp^3 hybridisation from arsenic is a reasonable description, which fits in nicely with the model of delocalisation via $\sigma-\pi$ overlap.

Yet again, β SiR_3 and SnR_3 groups have been found to exhibit the same preference for structure 8.3, and this seems to be a general rule. It is interesting that this preference has also been firmly inferred from certain kinetic data in which carbocations are implicated [8.12].

8.3 Sulphur-centred radicals and related species

This section is concerned primarily with $RS\cdot$ and $R_2S\cdot^+$ radicals, together with their 'dimers' $RS\dot{-}SR^-$ and $R_2S\dot{-}SR_2^+$. The corresponding selenium and tellurium radicals have not been widely studied.

e.s.r. studies of the very important thiyl radicals ($RS\cdot$) have been beset with errors, largely because satellite features from $R^{33}S\cdot$ radicals are almost always too weak to detect. Hence identification had to rest upon proton hyperfine coupling from the R-group and the form of the g-tensor components. Although this book is meant to constitute an overview of chemical applications, this area is examined in depth in order to give the reader some idea of the sort of problems that are frequently encountered. The technique of pulse radiolysis [8.13] has played a major rôle in this area, and it will be necessary to invoke results therefrom in the unfolding of the problem (see ref. 8.14 for a full account of this problem).

Disulphides, (RS—SR) are powerful electron scavengers:

$$RS\text{---}SR + e^- \rightarrow RS\dot{-}SR^- \qquad (8.3)$$

and in solid-state radiolyses, the parent anions can be detected by e.s.r. spectroscopy, and identified by the equivalence of the two R-groups. (For example, $MeS\dot{-}SMe^-$ gives a septet from the two methyl groups.) Liquid-phase studies revealed the formation of $RS\dot{-}SR^-$ radicals by their intense absorption band in the 420 ± 20 nm region (probably $\sigma \rightarrow \sigma^*$, as discussed below). They showed that, following reaction 8.4, dissociation to give RS· radicals was rapid:

$$RS \dot{-} SR^- \rightleftharpoons RS \cdot + RS^- \tag{8.4}$$

and reactions characteristic of RS· were detected. Some studies suggest that RS· radicals have a relatively weak absorption band in the 320 nm region. The same transient band at *ca.* 420 nm is observed when alkaline solutions of thiols are irradiated, showing that Equation 8.4 is reversible, and we have shown that the $RS \dot{-} SR^-$ e.s.r. spectrum results when alkaline glasses containing thiols are irradiated and annealed (the reverse of 8.4).

Thus, $RS \dot{-} SR^-$ anions are well established. However, RS· radicals are not. For many years, radicals having g-values close to 2.058, 2.025 and 2.001 (species X) were thought to be RS·. This is most unlikely, since these spectral features are almost independent of the environment, whereas those for RS· radicals are expected to be largely controlled by the environment, with g_z (z is along the C—S bond) $\gg g_x \sim g_y$. The detection of two different types of ^{33}S nuclei in species X confirms this, and shows that X must be formed from thiols by a bimolecular process [8.20]. My view is that species X has a structure analogous to that for $RS \dot{-} SR^-$, which clearly has its excess electron in the S—S σ^* orbital, i.e. $RS \dot{-} SR_2$, formed, in the case of thiols by

$$RS \cdot + RSH \rightleftharpoons RS \dot{-} S(H)R \tag{8.5}$$

Hadley and Gordy favour the simpler formulation RSS· [8.15]. There are strong arguments on both sides, outlined in ref. 8.19, and although my view, on chemical grounds, is that $RS \dot{-} SR_2$ is likely to prove to be correct, the problem remains open.

On the positive side, there can be little doubt that RS· radicals have now been properly detected by e.s.r. spectroscopy, both in crystals of various thiols, including cysteine ($\cdot SCH_2CH(NH_3^+)CO_2^-$) and penicillamine ($\cdot SC(Me)_2CH(NH_3^+)CO_2^-$), and also from a wide range of thiols in dilute solutions in CD_3OD. The major evidence in favour of this identification is the great variability of the $g_\parallel(g_z)$ values, which range from 2.158 for solutions in methanol glasses to 2.321 in crystalline cysteine hydrochloride, $\dot{S}CH(NH_3^+)CO_2H$.

The important point is that g_z is strongly dependent upon the medium, but not upon the nature of R. This is as expected because g_z depends on the way in which the medium (in this case, hydrogen-bonding) lifts the formal degeneracy of the $3p_x$ and $3p_y$ orbitals (see Section 2.1). In the presence of RS^- ions, these signals decay to give those characteristic of $RS \dot{-} SR^-$ ions, and in the presence of RSH they give species X—, possibly ($RS \dot{-} S(H)R$).

The marked tendency for RS· radicals to bond to sulphur is again displayed by $R_2S \cdot^+$ radicals. In fact, I know of no unambiguous evidence for $R_2S \cdot^+$ radicals, the many species claimed in the literature to be this cation being better identified as the 'dimer', $R_2S \dot{-} SR_2^+$. These cations have been well identified in the liquid phase by e.s.r. spectroscopy, an example being $Me_2S \dot{-} SMe_2^+$ which gives the expected set of 13 hyperfine components from four equivalent methyl groups [8.16]. They are characterised by an intense absorption in the 500–600 nm region.

Sulphur radicals of these types are thought to play several important rôles in biochemical processes, in particular in the interconversion of RSH and RSSR molecules. However, although the very extensive studies on such important molecules as cysteine and cystine mentioned above have shed light on the nature of the radicals involved, they have not yet been detected during normal *in vivo* reactions, so far as I am aware.

A variety of oxy-radicals of the type $R\dot{S}O$ and $R\dot{S}O_2$ have also been detected by e.s.r. spectroscopy. Also, a species having $g = 2.0096$ formed from various sulphur compounds in the liquid phase and often identified as the RS· radical is now thought to be an oxy-sulphuranyl radical, $R\dot{S}O(OR)_2$. In fact, RS· radicals have never been detected in the liquid phase, and probably never will be by conventional e.s.r. spectroscopy. This is because the precise hydrogen bonding needed to give a well defined solid-state spectrum will be subject to major and rapid fluctuations, including complete breaking, in the liquid phase, and this alone will suffice to spread the spectra over a very wide range. These fluctuations will also provide a ready route to rapid spin inversion and consequent line-broadening.

8.4 Phosphorus-centred radicals and related species

The three main phosphorus-centred organic radicals that have been detected and characterised by e.s.r. spectroscopy are $R_2P·$ (phosphinyl), $R_3P·^+$ (phosphoryl) and $R_4P·$ (phosphoranyl). The phosphinyl radicals (iso-electronic with $R_2S·^+$ mentioned in Section 8.3) are π-radicals like the amino radicals, the unpaired electron being essentially confined to a $3p$-orbital on phosphorus. A typical example, $Ph_2P·$ [8.17] prepared from Ph_2PCl by dissociative electron capture:

$$Ph_2PCl + e^- \rightarrow Ph_2P· + Cl^- \tag{8.6}$$

had $A_{\parallel}(^{31}P) = 268\,G$ and $A_{\perp} \approx (-)\,13\,G$, giving $A_{iso} = 78.7\,G$ and $2B = 189.3\,G$. The $2B$ value, compared with the $2B^0$ value of $201\,G$ shows that delocalisation onto the two aromatic rings is small. The small isotropic coupling confirms the π-nature of these radicals.

Phosphoryl radicals should have a σ-structure, being pyramidal at phosphorus, and this is confirmed by e.s.r. spectroscopy. The trialkyl derivatives are related directly to $·PH_3^+$, discussed in Chapter 7, and the trialkoxy derivatives, $·P(OR)_3^+$ and also $·P(O)(OR)_2$ and $·P(O)_2(OR)^-$, are related to $·PO_3^{2-}$, also discussed in Chapter 7. Various interesting trends in the ^{31}P hyperfine coupling constants have been discussed [8.18]. Some of these are illustrated in Fig. 8.7. On going from $·PR_3^+$ through to $·P(OR)_3^+$, two factors need to be considered. One is that pseudo π-delocalisation onto the $p(\pi)$ electrons of the oxygen ligands is expected to delocalise the unpaired electron and hence reduce the hyperfine coupling, and the other is that the increasing electronegativity of the ligands is expected to result in an increased $3p$-character from phosphorus in the σ-bonding orbitals, and a consequent increase in $3s$-character in the orbital of the unpaired electron. This opposition presumably explains the experimental curve shown in Fig. 8.7.

A better defined trend is observed for the phosphoranyl radicals on going from $·PR_4$ to $·P(OR)_4$ (also shown in Fig. 8.7). The parent radicals in this class

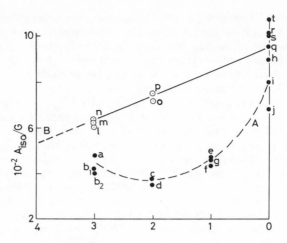

Figure 8.7 Some trends in the ^{31}P isotropic hyperfine coupling constants for various $\cdot PL_3$ (A) and $\cdot PL_4$ (B) radicals [8.18]. (a–j relate to PL_3 and l–t to PL_4 and AsL_4); (a) PH_3^+; (b_1, b_2) $PMe_3^+ + PEt_3^+$; (c) Ph_2PO; (d) Me_2PO; (e) HPO_2^-; (f) $PhPO_2^-$; (g) $PhPO(OH)$; (h) $P(OH)_3^+$: (i) $(MeO)_2\dot{P}O$; (j) $MeOPO_2^-$; (l) Me_3PO^-; (m) Me_3POR; (n) R_3AsOR; (o) $(RO)_2PR_2$; (p) $(RO)_2AsMe_2$; (q) $P(OMe)_4$; (r) AsO_4^{4-}; (s) PO_4^{4-}; (t) $As(OH)_4$.

are $\cdot PH_4$ and $\cdot PO_4^{4-}$, also mentioned in Chapter 7. It seems that electronegativity effects are of dominating importance, but σ-delocalisation onto the 'trans' ligands also plays an important rôle.

Many other phosphoranyl radicals are known with, for example, halogen, sulphur and amino ligands. Some reactions of these important radicals are considered in Section 10.2.2.

Various arsenic-centred radicals ($R_2As\cdot$, $R_3As\cdot^+$, $R_4As\cdot^+$, etc.) have also been quite widely studied. The results, including hyperfine coupling to ^{75}As ($I = \frac{3}{2}$) are remarkably similar to those for the phosphorus radicals discussed above. (Some examples are included in Fig. 8.7.)

8.5 Silicon-centred radicals and related species

By far the most important radicals in this class are the $\cdot SiR_3$ radicals of which the silyl radical, $\cdot SiH_3$, is the parent. Just as $\cdot NH_3^+$ is planar, but $\cdot PH_3^+$ pyramidal, so also, $\cdot CH_3$ is planar, but $\cdot SiH_3$ is pyramidal (see Chapter 7). After some controversy, it now seems that $\cdot SiR_3$ radicals in general have about the same value for $A_{iso}(^{29}Si)$ as that for $\cdot SiH_3$ (*ca.* 180 G). This is probably coincidental, since there is a definite increase in $A_{iso}(^{27}Al)$ on going from $\cdot AlH_3^-$ to $\cdot AlR_3^-$, but a comparable fall on going from $\cdot PH_3^+$ to $\cdot PR_3^+$, [8.19]. The data suggest that alkylation makes the aluminium radicals become more bent and the phosphorus radicals become more planar. The changes have been discussed in terms of hyperconjugative delocalisation onto the alkyl groups. Consider the sequence $\cdot AlMe_3^-$, $\cdot SiMe_3$ and $\cdot PMe_3^+$. As the normal flattening occurs through this series (assigned to electronegativity differences between the

80

methyl groups and the central atoms), so $\sigma-\pi$ type overlap with the C—H bonds giving rise to hyperconjugative delocalisation will increase. Even more important is the effect of charge. This form of delocalisation increases rapidly with an increase in positive charge on the central atom (cf. Chapter 6) and hence delocalisation will be trivial for $\cdot AlMe_3^-$ but marked for $\cdot PMe_3^+$. Hence an increase in flattening is to be expected for the alkyl derivatives of $\cdot PH_3^+$.

Similar results have been obtained for $\cdot SnR_3$ and $\cdot PbR_3$ radicals, both of which are again clearly pyramidal as judged by A_{iso} (^{117}Sn and ^{119}Sn) and (^{207}Pb) [8.20].

A variety of other organo-tin and lead radicals have been reported. For example, an interesting species formed in irradiated lead(II) acetate, had a large, nearly isotropic coupling to ^{207}Pb, and coupling to three equivalent protons. It was suggested [8.21] that the species was $Pb\dot{-}CH_3^{2+}$ with the unpaired electron in the Pb—C σ^* orbital, the —CH_3 group retaining its normal pyramidal structure. This evidence, that an alkyl radical can add to an ion such as Pb^{2+} is of some importance, and ties in well with the chemical properties of such systems.

8.6 Other systems

Most of the radicals discussed above would normally be considered to fall into the province of the organic chemist. Several other radicals have been described that have organic adenda, but fall, perhaps, closer to the inorganic chemist. For example, the radicals $R\dot{H}g\cdot$ and $R\dot{H}gR^-$ have been detected in irradiated alkylmercury(II) halides, RHg—hal, and dimethylmercury [8.22]. RHg· radicals are isostructural with the $\cdot PbMe^{2+}$ species discussed in Section 8.5, the extra electron being accommodated in the metal-carbon σ^* orbital. Somewhat surprisingly, the magnitude of the metal hyperfine coupling leads to the conclusion that the spin-densities in the $6s$ orbitals on Hg and Pb are comparable. This probably means that the orbital is close to being non-bonding rather than antibonding, since in the latter case the spin-density on mercury should be much greater than that on lead.

Another class of radicals which could properly be listed in this chapter can be described as transition-metal complexes with radical ligands. One example must suffice: this can be viewed as a nitroxide radical, $RR—\dot{N}O$, in which one of the R-groups is a transition-metal complex, say, $L_5M—$, as suggested by Swanwick and Waters [8.23]. These complexes are, like normal nitroxides, reasonably stable, and can often be studied in the liquid-phase. Such complexes are discussed further in Section 12.3.2.

CHAPTER 9

Environmental effects

The technique of e.s.r. spectroscopy has taken its place amongst other forms of spectroscopy as a useful method for obtaining structural and dynamic information about solvation, despite the fact that it is limited to the study of dilute paramagnetic solutes. This is especially true in the field of ion-pair formation, in which e.s.r. spectroscopy stands out as perhaps the most powerful technique for learning about their presence, structure and energetics [9.1–9.3]. It is also of considerable use for probing dynamic processes in which different species have life-times in the region of 10^{-6}–10^{-9} s, which is a range that is not readily accessible by other techniques.

9.1 Ion-pairs

Although it was appreciated that ion-pairs could have a range of possible structures, very little direct information about such structures existed before the discovery by Weissman and his co-workers [9.4] that the e.s.r. spectra of certain radical-anions in low polarity solvents contained hyperfine features characteristic of the alkali-metal gegen ions. These were described as ion-pairs rather than molecules because the anions retained their symmetry, as indicated by their e.s.r. spectra, and hence specific σ-bonding between the metal and a particular atom of the anion was ruled out. Furthermore, the addition of a more polar solvent led to the appearance of e.s.r. features of the normal anion with no contribution from the gegen ions. Further studies have confirmed and extended the original observations [9.1,9.2], and a remarkably detailed insight into the nature of ion-pairs has been forthcoming.

9.1.1 *Radical anions: cation hyperfine coupling*

The reason why alkali-metal nuclei frequently give rise to detectable hyperfine coupling is because slight electron transfer from the paramagnetic anions, which are powerful electron donors, takes place into the outer s-orbital of the cation, and hence even a small transfer gives an easily detectable isotropic coupling. For example, for the Na^+-benzophenone anion pair, a splitting of

1.125 G corresponds to *ca.* 0.36% transfer. This is typical, although values as great as *ca.* 4% transfer have been detected. Various detailed mechanisms for spin-transfer have been proposed [9.2,9.3]. One important factor is the extent of overlap facilitating charge transfer. If the preferred site for the cation is at or near a node in the unpaired electron's wave function, direct transfer is inhibited. Another of the factors that will control the extent of spin-transfer for a given pair of ions is the extent to which they are solvated, since strong anion solvation will hold back the electron and strong cation solvation will block its arrival as well as tending to separate the ions. Once a true 'solvent-shared' ion-pair has been formed, electron transfer will be prevented, at least for most insulating solvent molecules. The presence of such ion pairs may still give rise to a detectable modification of the e.s.r. spectrum of the anions but this is not such an immediately compelling effect as the appearance of metal hyperfine coupling.

For systems whose e.s.r. spectra display metal-ion hyperfine coupling, a variety of equilibria can be studied and often their rates and even their activation energies and entropies can be estimated from line-width effects over a range of temperatures. Examples of such equilibria are:

$$M^+ + A^- \rightleftharpoons (M^+A^-)_1 \rightleftharpoons (M^+A^-)_2 \rightleftharpoons (M^+A^-)_c \qquad (9.1)$$

$$M^+ + A^-M^+ \rightleftharpoons M^+A^- + M^+ \qquad (9.2)$$

$$M^+ + A^-M^+ \rightleftharpoons M^+A^-M^+ \qquad (9.3)$$

$$M^+A^- \rightleftharpoons A^-M^+ \qquad (9.4)$$

$$A + M^+A^- \rightleftharpoons A^-M^+ + A \qquad (9.5)$$

$$A^- + M^+A^- \rightleftharpoons A^-M^+ + A^- \qquad (9.6)$$

$$A^- + M^+A \rightleftharpoons A^-M^+A^- \qquad (9.7)$$

In Process 9.1 $(M^+A^-)_1$ represents the ion-pair with maximum solvation, the limiting form being the contact ion-pair $(M^+A^-)_c$. If the reactions interconverting successive ion-pairs $(M^+A^-)_i \rightleftharpoons (M^+A^-)_j$ are fast on the e.s.r. time-scale (see Chapter 5), then a single averaged spectrum will result. Alternatively if these equilibria are relatively slow, the contact ion-pair will be well defined, but it may be difficult to distinguish between other ion-pairs and the solvated anions. I stress, however, that the extrusion of solvent molecules on ion-pairing need not always occur as a series of well defined jumps. Thus one can envisage a more continuous ebb and flow process such as that in Scheme 9.8 with no specific route involved. Apart from the contact ion-pair, the best defined unit is likely to be that solvent shared ion-pair in which the shared solvent molecule is amphoteric, but even here, a range of possibilities exist, because of the non-linearity of the bonding functions.

(9.8)

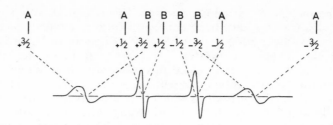

Figure 9.1 Predicted hyperfine features for a cation having $I = \frac{3}{2}$ in equilibrium between two different types of ion-pair, A and B.

Evidence that different types of ion-pairs are sometimes involved is strong [9.2]. For example, Hirota [9.5] found that lithium anthracenide in diethyl ether gave spectral features for the normal ion-pair showing cation hyperfine coupling together with a species showing no cation coupling, but having proton splittings that were different from the normal solvated ion values.

In the fast-exchange region, the e.s.r. spectra for two types of ion-pair with different cation hyperfine coupling would be expected to exhibit selective broadening of the individual features (B-N-N-B) as depicted in Fig. 9.1. (Differences in the $\langle g \rangle$ and $\langle A \rangle$ values for the anions would make this an asymmetric effect.) Such effects have been observed for several systems.

In these studies, especially those involving a second solvent such as the chelating glymes, the two types of ion pair clearly differ in the extent of solvation. A different situation arises when there are two different 'binding' sites in the anion. Again, two different types of ion-pair should be detectable in the slow-exchange regime and an average with suitable line-broadening in the fast-exchange region. The latter situation was observed for 2,6-dimethyl-p-benzosemiquinone [9.6], there being a clear preference for the less hindered carbonyl oxygen.

Process 9.2, the bimolecular analogue of Process 9.1, is readily detectable because it causes an uncertainty in the nuclear quantum number for M^+. This leads to broadening and, ultimately, loss of hyperfine structure, the resulting spectrum in the fast exchange region being almost indistinguishable from that of the solvated anions alone. The linewidth increments lead directly to rates.

9.1.2 Anions with two binding sites

However, if the anion has more than one equivalent binding site, as for the semi-quinones or dinitrobenzene anions, Process 9.3 may occur and the triple ion may have sufficient stability to give rise to separate resolved features including hyperfine components characteristic of two equivalent cations [9.7,9.8]. (See Section 9.1.5). As expected, the spectral features for the anions within such triple ions are once more typical of the normal, symmetrical species, whereas for the ion-pair they reflect the asymmetry induced by the single perturbing cations.

Equilibrium 9.4 is another manifestation of such asymmetry. When there are two equivalent sites, the cation can migrate from one to the other at a rate sufficiently great to modify the e.s.r. spectrum. Typically, at low temperature

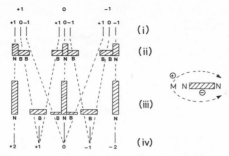

Figure 9.2 Hyperfine features for a radical anion with two equivalent ^{14}N nuclei ($I = 1$). (i) On association with a cation which induces a fluctuation in the ^{14}N hyperfine coupling the coupling constants of the two ^{14}N become different. (ii) Slow exchange. (iii) Intermediate exchange. (iv) Fast exchange. (N = narrow; B = broad) If exchange occurs at a slow rate this will broaden as in (ii), and if it is fast they will exhibit alternate broadening as in (iii).

the spectrum may be that of the asymmetric anion, whilst that at ambient temperature may be that of the symmetric anion. At intermediate temperatures selective line broadening occurs because only alternate lines are different for the two limits (see Fig. 9.2). From this broadening the rate of migration can be calculated. The spectra reveal that it is a single cation that is responsible for the spectral changes, since its contribution to the hyperfine coupling is unaffected.

One advantage of the e.s.r. method is that many processes occur at rates that strongly affect linewidths. This is true for most intramolecular cation migrations [9.2]. One example that has been very thoroughly studied is that of the anthraquinone radical anion [9.9]. Values for $\Delta G^{0\neq}$, $\Delta H^{0\neq}$ and $\Delta S^{0\neq}$ have been obtained for sodium and potassium ion migrations in THF over a wide temperature range, and the effect of adding DMF to these solutions has also been examined. The overall results are:

(i) Activation energies are small (3–4 kcal mol^{-1}), with values for K$^+$ slightly larger than those for Na$^+$.

(ii) $\Delta S^{0\neq}$ values are relatively large and negative, with those for Na$^+$ (*ca.* -17 e.u.) greater in magnitude than for K$^+$ (*ca.* -9 e.u.).

(iii) Added DMF reduces the magnitude of $\Delta S^{0\neq}$, resulting in a marked increase in migration rate.

The key factor is the large negative entropy, which for this solvent, which bonds strongly to the cations only, means that the cations are more strongly solvated in the transition state. This is to be expected since the anion-cation bond has to be severed cleanly prior to the making of the new bond at the second oxygen site. Indeed, a reasonable model might well be one of displacement by an incoming solvent followed by migration and subsequent replacement of a solvent molecule by the second oxygen (Scheme 9.9). Since solvent is bound more weakly to K$^+$ than to Na$^+$ $|\Delta S^{0\neq}|$ is reduced. The net contribution to $\Delta H^{0\neq}$ is then slightly greater on balance for K$^+$.

$$\text{S} \quad \text{O—C} \rightleftharpoons \text{C—O} \rightleftharpoons \text{O—C} \quad \text{C—O} \rightleftharpoons \text{O—C} \quad \text{C—O} \quad \text{S} \tag{9.9}$$

The effect of added DMF must also be understood in terms of cation solvation. This is a stronger cation solvator than THF and hence even at low concentrations will be preferentially bonded to the cations in the ion-pairs. This will have the effect of weakening the cation-anion interaction and reducing the need for extra solvation in the transition state.

9.1.3 *Solvation versus ion-pair formation*

The main factor involved in these studies of ion-pairs is solvation, upon which the electrostatic attraction of cations and anions is superimposed. It is important to realise that the highly localised hydrogen bonds between solvent molecules such as alcohols or amides and an anion may be more effective in stabilising (and possibly perturbing) that anion than a cation such as Na^+ or even Li^+. Protic media are generally also basic media, and thus both cations and anions are strongly solvated. Solvent shared ion-pairs are quite acceptable since neither ion is desolvated, but contact ion-pairs are improbable except in very concentrated solutions, or when both anion and cation have a low surface charge density. Aprotic solvents, however, do not bind strongly to anions, so an anion can readily replace a solvent molecule as a ligand on the cation, the strength of bonding of the solvent and anion being comparable in many cases.

These considerations are illustrated by results for various nitroaromatic anions [9.10]. The hyperfine coupling to ^{14}N is sensitive to environment. Thus bonding of cations or protic solvent molecules to the oxygen atoms pulls negative charge onto oxygen, increases the spin-density on nitrogen and hence increases $A(^{14}N)$. A rough guide to the results is given for $PhNO_2^-$ in Scheme 9.10. The results suggest that the perturbing power of the alcohols is greater even than that of Li^+. This arises, presumably, because several alcohol molecules contribute to the effect. As expected, $Li^+ > Na^+ > K^+ > Rb^+ > Cs^+$ for ethereal solutions, and these all perturb the anions more than R_4N^+ ions in ethers. Indeed this value is probably close to the limit for unperturbed anions.

SOLVENT	R_2O	R_2O	R_2O	EtOH	R_2O	EtOH	EtOH		$PhNO_2^-(^{14}N)$
CATION	R_4N^+	Cs^+	Na^+	R_4N^+	Li^+	Nil	Na^+		(9.10)
	10		11		12		13	14	(Gauss)

Addition of R_4N^+ salts to the alcohol solvate results in strong anion solvation and consequent competitive desolvation of $PhNO_2^-$ ions, causing a steady fall in $a(^{14}N)$. In contrast sodium ions cause a small increase in $a(^{14}N)$. This, we suggested, means that solvent shared ion-pairs cause a slightly greater perturbation by strengthening one of the hydrogen bonds to oxygen [9.10].

9.1.4 *Solvation of ion-pairs*

Quite apart from the possibility of different types of ion-pair (contact, solvent-shared, etc.), we expect the e.s.r. parameters of ion-pairs, M^+A^-, to be solvent

86

sensitive. We have explored this possibility using various substituted semi-quinones in a range of solvents [9.11]. Several generalisations were extracted from the data.

(i) There was a steady fall in calculated spin-density on the cations on going from Li^+ to Cs^+, for all solvents. This is normal and reflects the electron affinity changes or surface charge-density changes.

(ii) The hyperfine coupling to M^+ fell on cooling. This again is normal and is explained in terms of a preferential binding site for M^+ on or near a nodal surface for the unpaired electron, with increasing migration from this site with increasing temperature:

$$\text{C—O} \cdots\cdots (M^+) \text{ — NODAL PLANE} \qquad (9.11)$$

(iii) $A_{iso}(M^+)$ fell on going from durosemiquinone to the unsubstituted anion. This, we suggest, arises because the in-plane site is sterically difficult or even impossible to occupy in the methylated anion, thus forcing the cation into the out-of-plane site which gives better overlap and more transfer of spin (as in Scheme 9.11).

(iv) The metal hyperfine coupling fell on going from t-pentanol to ethereal solvents. (Other alcohols gave dissociation of the ion-pairs, or solvent shared ion-pairs.) This is, at first sight, surprising, since solvent shared ion-pairs would be expected and these should give small or zero metal coupling. However, in view of the bulky nature of the solvent, such structures may be improbable. We suggest that solvation occurs at the carbonyl site not occupied by the cation, thus minimising the perturbation and increasing the spin-density on the carbonyl group near the cation. This is supported by the proton hyperfine coupling which moves closer to the unperturbed values in the alcoholic media.

It is interesting to compare these results with those of Nakamura *et al.* [9.12], who studied the e.s.r. spectra of alkali salts of fluorenone in mixed protic and aprotic media. The ^{13}C hyperfine coupling for the carbonyl carbon increased rapidly for all except the Li^+ salt on the addition of alcohol to ethereal solutions, the increases falling in the order MeOH, EtOH, iPrOH. The form of these curves suggests that one solvent molecule is initially involved, and the increase in $a(^{13}C)$ shows that this is hydrogen bonded to the carbonyl oxygen atom. In order for this to occur, some movement of the cation out of the radical plane is required for steric reasons, and this is indeed proven by the initial increase in $a(M^+)$.

In another interesting study of the solvation of ion-pairs, Chen *et al.* [9.13] added DMF to various alkali metal aromatic ketyls in ethers. Invariably, $a(M^+)$ increased as one DMF molecule became bound to the complex. Further solvation resulted in dissociation, in a one-step process, not involving solvent-separated or shared ion-pairs. As stressed above, this is expected for aprotic media since, once a solvent molecule has been inserted between the ions, there is no bonding left to hold them together other than weakened electrostatic forces. However, that $a(M^+)$ should increase prior to dissociation is not an obvious result. Since one DMF molecule has become bonded to the cation, and since this interaction is relatively strong, the anion–cation interaction must be weakened. In that case, thermal migration away from the in-plane site would be greater, and hence the coupling could increase, as observed. This result fits in with the increased rate of migration when DMF is added, that was discussed in 6.1(b).

9.1.5 *Triple ions and ion clusters*

Solutions of electrolytes in low-dielectric solvents tend to form ion aggregates even at low concentrations. e.s.r. spectroscopy is capable of defining at least two such species clearly. Thus $M^+A^-M^+$ ions, if they have a sufficiently long life, will show hyperfine coupling to two equivalent cations, whilst $A^-M^+A^-$ ions will usually be in a triplet-state which also has a characteristic spectrum (see Chapter 11).

The former species were discovered by Gough and co-workers [9.7,9.8] for anions having two equivalent functional groups. Their success came from the use of sodium tetraphenylboride. For this salt there can be no bonding between the cations and anions, so the possibility of bonding to the vacant sites of the radical anions becomes attractive and does in fact lead to stable triple ions.

As with much of the early work on ion-pairs, Weissman and co-workers pioneered the study of triplet-state aggregates by studying the liquid- and solid-state e.s.r. spectra of species containing divalent cations, $A^-M^{2+}A^-$ [9.14]. For monovalent cation salts, the species detected is $(M^+A^-)_2$ rather than $A^-M^+A^-$, as evidenced by hyperfine coupling from two equivalent cations [9.15]. Solid-state spectra now give the 'mean' separation between the interacting electrons, and some measure of any deviations from axial symmetry [9.2].

9.1.6 *Radical cations in ion-pairs*

Far more attention has been paid to the solvation and ion-pairing of radical anions than to that of radical cations. This is partly because media such as sulphuric acid were traditionally used. Also, as was stressed some time ago [9.1], slight electron transfer towards cations such as Na^+ results in slight population of their outer s-orbitals, giving rise to a readily detectable isotropic hyperfine coupling. However for $\cdot A^+X^-$ ion-pairs, electron transfer from X^- leaves unpaired spin in the p-manifold of X^-. This will not, in first order, give rise to any isotropic hyperfine coupling, which will therefore probably remain undetectably small. However, such a charge-transfer configuration will give rise to a field induced orbital angular momentum and this will in turn give rise to a small positive g-shift. These predictions have been nicely confirmed in recent work by Goez-Morales and Sullivan [9.16], who studied the cation of 1,2,4,5-tetramethoxybenzene in a range of solvents including chloroform, methylene chloride, methylcyanide, nitromethane and their higher alkyl analogues. The anion induced g-shifts were much larger for I^- (or I_3^-) than for Br^- (or Br_3^-), as expected, and became greater as the solvent polarity was reduced. These experiments clearly establish the presence of halide ion-pairs, but no further details about these species are available at present.

9.2 Solvation of radical anions

By far the most significant change in the e.s.r. spectra of radical anions occurs when solvent molecules can form hydrogen bonds to some specific atom or group. Thus addition of a protic solvent such as an alcohol to a solution of the radical anion in an aprotic solvent has little effect on the e.s.r. spectra of species such as the naphthalene anions, but a marked effect on the spectra of semiquinone anions [9.17] and nitro-aromatic anions [9.10]. Indeed these solvent induced perturbations may be as great or greater than those induced by cations in ion-pairs [9.10]. Most studies have been devoted to p-benzosemiquinones and aromatic nitro anions.

9.2.1 *Semiquinones*

Two extremes of solvation can be envisaged. In one, the solvent interacts to stabilise the negative charge equally on the two carbonyl groups, whilst in the other strong solvation at one carbonyl group pulls the negative charge predominantly to this end of the molecule, leaving the other side largely unsolvated. The former is more likely, and in fact solvent induced asymmetry has not been detected for semiquinones. If one considers an isolated anionic carbonyl group, $R_2C\dot{=}O^-$, then hydrogen bonding to oxygen will tend to stabilise the form $R_2\dot{C}-O^- \ldots HA$, thus forcing the unpaired electron more onto carbon. This should result in an increase in $a(^{13}C)$ and a fall in $a(^{17}O)$. For the semiquinones this occurs at both carbonyl groups simultaneously, and the expected changes in the coupling to ^{13}C and ^{17}O are indeed observed [9.18]. The ring proton hyperfine coupling is effectively 'buffered' and is relatively insensitive to such changes. Values of a_{iso} (^{13}C) or (^{17}O) are a linear function of the Z-value [9.19] of a range of solvents for p-benzosemiquinone [9.20] (Fig. 9.3). This means, indeed, that protic solvents are very much more effective in perturbing the anions, water being the most powerful. The reason why such good correlations with Z are observed (Fig. 9.3) is probably because in both studies, it is the

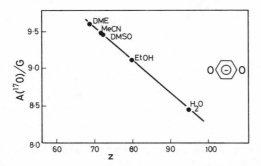

Figure 9.3 Dependence of the ^{17}O coupling constant of p-benzosemiquinone upon the Z value of the solvent.

tendency to pull negative charge towards the solvent molecules that is measured.

9.2.2 *Aromatic nitro-anions*

Results already discussed show that solvation of aromatic nitro-anions in aprotic solvents is weak and causes little perturbation. However, protic solvents form hydrogen bonds to the oxygen atoms and these are of sufficient strength to perturb the anions in a readily detectable way. When two functional groups are present, they are generally solvated equivalently. However, this is not the case in the special instance of *m*-dinitrobenzene anions [9.21].

In aprotic solvents, these anions are symmetrical. However, when ion-pairs are formed they become asymmetric in a catastrophic fashion [9.22], such that $a(^{14}N_1) \gg a(^{14}N_2) \approx 0$. In a crude sense this can be pictured as a switch to the mononitro structure, the other NO_2 group behaving as a neutral substituent. In alcoholic media, a linewidth alternation is observed even though ion-pairing is insignificant. This must mean that there is a rapid switching from strong to weak solvation at both nitro groups occurring out of phase. In the unique case of water, however, the slow exchange situation is found, with $a(^{14}N_1) \gg a(^{14}N_2)$. We deduced from the linewidths obtained from such spectra that the mean lifetime of these asymmetric solvates is *ca.* 4.5 μs at 282 K. Addition of *t*-butyl alcohol caused an initial increase in this lifetime followed, in the 0.04 mole fraction region by a rapid fall. This is, in my view, one of the most dramatic examples of the special rôle of water structure yet discovered.

9.3 Solvation of neutral radicals

Although stable nitroxide radicals have been widely exploited as 'spin-labels' in biological systems (Chapter 13), their use as probes to study solvation has not

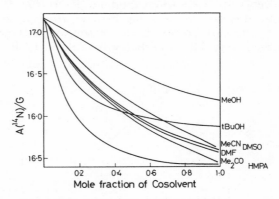

Figure 9.4 Hyperfine coupling constants (^{14}N) for DTBN in aqueous solutions as a function of the mole fraction of added co-solvents. (MeOH = methanol, *t*-BuOH = *t*-butyl alcohol, MeCN = methylcyanide, DMSO = dimethylsulphoxide, Me_2CO = acetone, HMPA = hexamethylphosphoramide and DMF = dimethylformamide.)

been extensive. Jolicoeur and Friedman [9.23,9.24] have reported the effect of varying solvent upon the two nitroxides usually known as TEMPO (2,2,6,6-tetramethyl-piperidine-N-oxide) and OTEMPO (2,2,6,6-tetramethyl-4-piperidone-N-oxide). They concentrated upon linewidth effects, distinguishing between a uniform broadening induced by spin-rotation relaxation (Y_J) and an asymmetric broadening ascribed to modulation of the g and A tensor components by the rotational motion (Y_c) (see Chapter 5).

We have carried out similar studies using di-t-butyl nitroxide (DTBN), but have focused attention primarily on changes in $A(^{14}N)$ [9.25]. Linewidth studies are rendered difficult by the fact that unresolved hyperfine coupling to many protons makes a major contribution for all but the most viscous solvents or highest temperatures. The difficulty is that $A(^1H)$ also varies with solvation, increasing in magnitude as $A(^{14}N)$ falls.

Some trends in $A(^{14}N)$ are displayed in Fig. 9.4. All these changes can be understood in terms of the hydrogen-bonding equilibrium 9.12:

$$R_2N\dot{O}\text{------}H\dot{O}H\text{-----} + B \rightleftharpoons R_2N\dot{O} + B\text{-----}H\dot{O}H\text{-----} \quad (9.12)$$

which is rapidly established on the e.s.r. time-scale for all fluid solutions.

$A(^{14}N)$ is a maximum in aqueous solutions, for which this equilibrium is assumed to be constrained completely to the left. Interaction with aprotic solvents is dipolar in nature and $A(^{14}N)$ is then relatively insensitive to changes. The major difference between $R_2N\dot{O}$ as a probe and anions such as $Ph\dot{N}O_2^-$ is its low basicity. Thus as [B] increases hydrogen bonds are rapidly lost. This could be viewed as preferential solvation by the aprotic solvent, but in our view this is a completely misleading description. It is the strong affinity of the basic aprotic solvent for hydrogen bonds that removes water from the nitroxide, and it is the loss of such bonding, not the gain of aprotic solvent molecules, that causes $A(^{14}N)$ to fall.

The trend to lower $A(^{14}N)$ values on adding methanol could be caused by weaker H-bonding, or by an increase in the concentration of non-H-bonded nitroxide. Since methanol is a relatively strong proton donor, we favour the latter explanation, and this is supported by recent solid-state studies which suggest the presence of two types of nitroxide in methanol glasses at 77 K [9.25].

The trend for t-butyl alcohol is particularly interesting. There is an initial plateau in which $A(^{14}N)$ doesn't change appreciably, but this is progressively lost on heating. Then there is a rapid fall in $A(^{14}N)$, the limiting value being nearly reached in the 0.4 M.F. region. The plateau and its temperature sensitivity is thought to be characteristic of the unique ability of water to form cages about large spherical hydrocarbon groups [9.26]. Such cage formation is typified by results for aqueous t-butyl alcohol, and is expected for DTBN. Thus, until there is insufficient water for these cages, the nitroxide is not greatly affected. However, once this region (*ca.* 0.04 M.F.) is passed, cage sharing is thought to set in, and this is reflected in the rapid fall in $A(^{14}N)$. Dispersion of the cages leaves the residual water preferentially solvating the alcohol molecules (9.13).

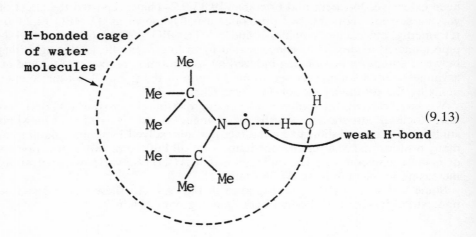

(9.13)

9.4 Information from radiation studies

Cation-anion interactions of the type observed with ion-pairs discussed above is frequently detected for radicals formed in ionic solids by ionising radiation [9.2,9.27,9.28]. These studies have helped in the task of deducing details of geometry for ion-pairs since for single crystal studies radical orientations are frequently known. Another significant way in which one can learn about matrix effects is by studying a given radical in a range of different salts and solvents. Neutral molecules interact only weakly with their environment, whereas radical anions such as CO_2^-, SO_3^- or PO_3^{2-} interact strongly, their e.s.r. parameters $(A_{iso}, 2B)$ being strongly dependent upon the surrounding crystal field [9.29].

CHAPTER 10

Some aspects of mechanism

Long before the advent of e.s.r. spectroscopy it was realised that many liquid-phase reactions proceeded via radical intermediates of short life. Detailed mechanisms were devised to explain the range of products and the overall kinetic course of the reactions. A variety of tests for radical intermediates included the effect of oxygen (which is an efficient radical scavenger) and of certain vinyl monomers, such as acrylonitrile ($H_2C\!=\!CHCN$), which form polymers with very high efficiency in the presence of trace concentrations of radicals.

e.s.r. spectroscopy has extended mechanistic studies because radical intermediates can now be directly detected and identified, and the intermediate rates of decay and interconversion of radicals can be measured.

I should stress that two other techniques are now also used extensively in the study of radical mechanisms—namely ultraviolet spectroscopy and n.m.r. spectroscopy (using the phenomenon of nuclear spin polarisation, CIDNP). The former technique is used in conjunction with an intense flash of light in the technique known as 'flash-photolysis' [10.1], or a pulse of ionising radiation in the technique known as 'pulse-radiolysis' [10.2]. It has the great advantage, relative to e.s.r. spectroscopy, that very short time intervals (in the region of picoseconds using sophisticated laser techniques) can be studied, but the disadvantage that optical spectra are often very broad, and may lack the detail needed to give firm identification. This means that arguments based on kinetic studies, product analysis and simply chemical expectation have often to be used to identify a given intermediate.

The n.m.r. technique CIDNP (chemically induced dynamic nuclear polarisation) is extremely useful in demonstrating which products have been formed by a radical process, since different components of their n.m.r. spectra may, under certain circumstances, have markedly different intensities from normal, including negative intensities (i.e. in emission rather than absorption). The phenomenon of CIDNP which is of great significance in free-radical chemistry, is reviewed, for example, in ref. 10.3. It is closely related to the e.s.r. analogue, CIDEP, and both are discussed in Section 11.6.

Before discussing some examples of the way e.s.r. spectroscopy has been used to probe mechanistic detail, an outline of its use in probing various methods of radical generation is presented.

10.1 Generation of radicals

10.1.1 *Photolysis*

This is probably the most common and versatile method. Using a microwave cavity fitted with a grating designed to let light onto the cell, but not to perturb the microwaves, *in situ* photolysis becomes quite simple, and is very effective because quite high stationary concentrations of radicals can be accumulated even when they are lost by diffusion controlled radical-radical reactions. (For typical generation rates, life-times are of the order of milliseconds.) Sometimes the experiments are designed to study intermediates formed in the photolysis of a molecule of interest, but more frequently a photoactive molecule, often a peroxide, is used as a sensitisor, to initiate reactions of interest. For example, Livingston and his co-workers have used solutions of hydrogen peroxide to generate ·OH radicals, which, though not themselves detected, go on to react with various substrates, usually by addition or hydrogen atom abstraction, to give detectable radicals [10.4]

$$H_2O_2 \xrightarrow{h\nu} 2 \cdot OH \tag{10.1}$$

$$\cdot OH + RH \rightarrow H_2O + R \cdot \tag{10.2}$$

We used this process many years ago to generate a range of radicals, but in those days it was necessary to use glasses at low temperature in order to accumulate enough radicals for detection by the relatively insensitive instruments then available [10.5]. In fact, although the hyperfine features were greatly broadened by the dipolar hyperfine coupling, the larger splittings were well defined, and in most instances correct identifications were made. These studies have confirmed, for example, that alcohols tend to loose C—H hydrogen atoms adjacent to the hydroxyl group:

$$R_2CHOH + \cdot OH \rightarrow R_2\dot{C}OH + H_2O \tag{10.3}$$

A knowledge of the reactivity of ·OH radicals is of importance since this is a major primary product in the radiolysis of aqueous systems.

This general method has the advantage that most media can be used. For non-aqueous systems, di-*t*-butyl peroxide is a popular sensitisor. In addition to being a source of $Me_3C\dot{O}$ radicals, it is also a source of methyl:

$$(CH_3)_3C\dot{O} \rightarrow \cdot CH_3 + (CH_3)_2CO \tag{10.4}$$

which, in contrast with $(CH_3)_3CO\cdot$ radicals, can be detected directly by e.s.r. spectroscopy in the liquid phase.

Another important method of generating radicals by photolysis is via excited triplet-states (Chapter 11). Triplet molecules can often be thought to have reactivities that resemble a composite of the two electrons involved in the triplet. Thus, in the charge-transfer system the reactivity will, in some respects, resemble that of the electron-loss species, $\bigcirc_A\bigcirc_B^+$, combined with that of the electron-gain species, $\bigcirc_A\bigcirc_B^-$. For example, triplet aldehydes or ketones can extract reactive hydrogen atoms from molecules in the same way as alkoxy radicals:

$$R_2\dot{C}\!-\!\dot{O}(t) + R'H \rightarrow R_2\dot{C}OH + R\cdot' \qquad (10.5)$$

10.1.2 Radiolysis

Whereas near-ultraviolet photolysis leads generally to bond homolysis, radiolysis results, initially, in electron ejection:

$$AB \xrightarrow{h\nu} AB^+ + e^- \qquad (10.6)$$

Low-temperature studies of solids (crystals or glassy solutions) have frequently revealed the presence of primary AB^+ 'electron-loss' centres, but these often go on to give secondary radical products. Ejected electrons may become physically trapped (or solvated), or they may react directly with AB (or with added substrates) to give 'electron-gain' centres:

$$e^- + AB \rightarrow AB^- \qquad (10.7)$$

Some examples of such reactions are given in Section 10.2, below.

Liquid-phase radiolysis requires far more sophisticated equipment, since an *in-situ* method is necessary. Fessenden and Schuler [10.6] pioneered a steady-state method, in which a beam of high energy electrons is directed through a narrow hole in the centre of the pole-pieces of the magnet, into the cavity. Since then, many interesting studies have been forthcoming from Fessenden and his co-workers using steady-state and pulse techniques, but because of the difficulties involved the method has not been widely adopted. In addition to giving the structure or concentration of radical intermediates, it also gives kinetic data (Section 10.3) and often CIDEP information (Section 10.5).

10.1.3 Thermolysis

Although this is not widely used as a method of obtaining radicals of interest in the solid or liquid phases, it is used extensively in the gas phase, and this can then be combined with the technique of matrix-isolation to give radicals permanently trapped in the inert matrix. An example is the generation of $Hg(CN)$ from $Hg(CN)_2$ [10.7]:

$$Hg(CN)_2 \xrightarrow{heat} \cdot Hg(CN) + \cdot CN \qquad (10.8)$$

Other thermolyses, occurring at far lower temperatures, such as:

$$N_2O_4 \rightleftharpoons 2\cdot NO_2 \qquad (10.9)$$

$$N_2F_4 \rightleftharpoons 2\cdot NF_2 \qquad (10.10)$$

$$S_2O_4{}^{2-} \rightleftharpoons 2\cdot SO_2{}^- \qquad (10.11)$$

$$Ph_3C(H)\left\langle\!\!\bigcirc\!\!\right\rangle\!=\!CPh_2 \rightleftharpoons 2\cdot CPh_3 \qquad (10.12)$$

95

and

$$(R_2NO)_2 \rightleftharpoons 2R_2\dot{N}O \qquad (10.13)$$

are normally taken for granted since radical formation is extensive at room temperature. The reverse of such homolyses constitutes one important mode of radical loss, as discussed below in Section 10.3. The σ-bonds that are broken in Reactions 10.9–10.13 are weak for inherent electronic reasons, as evidenced by their unusual length. A steric contribution may also be important in preventing extensive dimerisation for radicals such as the nitroxides, R_2NO, which normally have bulky R-groups. This principle has been extensively utilised by Ingold and his co-workers in order to give a range of relatively stable radicals which, in the absence of bulky substituents, would readily dimerise [10.8]. Tertiary alkyl substituents are also important in preventing the occurrence of bimolecular disproportionation reactions involving abstraction of α-hydrogen atoms.

10.1.4 *Redox processes*

The radiolytic reactions discussed in Section 10.1.3 above really constitute examples or redox reactions. Most other methods, including electrolyses and chemical reactions, are very much more selective. Electrode reactions have been widely used to generate relatively stable electron-gain and electron-loss centres such as aromatic radical cations and anions. The method requires the presence of ions, so usually media such as the alcohols or dimethylformamide have been utilised.

Alkali metals, especially sodium, have been widely used as electron donors, either directly or via solvated electrons (Section 7.1):

$$Na \rightarrow Na^+_{solv} + e^-_{solv} \qquad (10.14)$$

$$e^-_{solv} + X \rightleftharpoons X^- \qquad (10.15)$$

Other less powerful and hence more selective electron donors include SO_2^- (from $Na_2S_2O_4$), $H_2CNO_2^-$ (from $MeNO_2$ in alkaline solution) and O_2^- (from NaO_2 or KO_2).

Simple electron acceptors are usually transition metal ions, Ce(IV) in aqueous solution being an important example. Other 'acceptors' are more 'chemical', in that other reactions besides electron transfer may result from their use. Examples include MnO_4^- (giving MnO_4^{2-} as one possible mode of reaction), $Pb(OCOMe)_4$, PbO_2, etc.

One important class or reactions that relate to these are based on Fenton's reagent, which comprises a mixture of ferrous ions and hydrogen peroxide in acidic aqueous solutions. This reagent reacts with a wide range of substrates by what are termed 'induced oxidations', which do not occur except in the presence of both components of the reagent. The simplified Haber–Weiss mechanism usually put forward to explain these reactions are:

$$Fe^{2+} + H_2O_2 \rightleftharpoons FeOH^{2+} + \cdot OH \qquad (10.16)$$

$$\cdot OH + R\!-\!H \rightarrow H_2O + R\cdot \qquad (10.17)$$

This particular reagent has not been very widely used with e.s.r. spectroscopy, but the comparable one using titanous ion has been used extensively, especially by Norman, Gilbert and their co-workers [10.9]. In order to catch the hydroxyl radicals before they react in other ways, it is necessary to use a flow system in which mixing occurs, either directly in the cell or just as the reagents enter the cell. As stressed repeatedly, hydroxyl radicals are not detected as such in the liquid-phase (Section 7.3) but their complexes with Ti(IV) are. (Narrow singlets at $g = 2.0128$ and 2.0114.)

It is important to note that the products or reaction are not precisely the same as those obtained under identical conditions when titanous ions are absent, and ultraviolet light is used to generate hydroxyl radicals. Thus, for example, ethanol gives a considerable yield of $H_2\dot{C}CH_2OH$ from the titanous-peroxide system, in addition to the $CH_3\dot{C}HOH$ radical, which is the only major product from peroxide photolysis. Also, allyl alcohol, $CH_2{=}CHCH_2OH$, which gave largely $CH_2{=}CH{-}\dot{C}HOH$ after photolysis, gave extensive addition as well, with the titanous reagent. This shows that the titanium ions play an active rôle in the reaction, which may well involve co-ordination of the substrate and the hydroxyl radicals to the metal ion. If this is the case, the reaction has strong links to the important reactions involving homogeneous catalysis by transition metal ions, many of which involve the formation of radical intermediates (see, for example, ref. 10.10).

10.2 The study of radical intermediates

This constitutes the most important aspect of the mechanistic application of e.s.r. spectroscopy, but care must be used to guard against the firm conclusion that a detected radical plays a major part in a given process. It may well be the product of a minor side reaction that is fortuitously more persistent than the significant intermediates. Generally, low-temperature solid-state studies do reveal primary or significant secondary radicals, but liquid-phase studies may give dominating signals from the most stable of a range of radical intermediates. If this happens to be formed by a minor pathway then incorrect conclusions can result.

Some examples of mechanism probed via the identification of significant intermediates are now outlined.

10.2.1 *Additions and β-eliminations*

Here we are concerned with the forward and reverse reactions:

$$X\cdot + R_2C{=}Y \rightleftharpoons R_2\dot{C}{-}Y{-}X \tag{10.18}$$

One example, involving the formation of methyl radicals from $(CH_3)_3C\dot{O}$, has already been mentioned in Section 10.1 (equation 10.4). Others include the following reactions, of importance in polymer chemistry [10.11].

$$\cdot OH + CH_2{=}CXY \rightleftharpoons HOCH_2{-}\dot{C}XY \tag{10.19}$$

$$HOCH_2{-}\dot{C}XY + CH_2{=}CXY \rightleftharpoons HOCH_2CXY{-}CH_2{-}\dot{C}XY \text{ (etc.)} \tag{10.20}$$

(where $X = H$, CH_3 and $Y = CO_2H$, CO_2Me, CN, etc.)

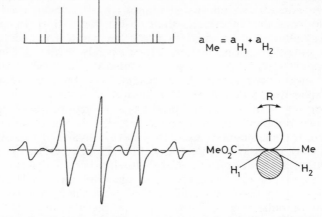

$$a_{Me} = a_{H_1} + a_{H_2}$$

Figure 10.1 First derivative X-band e.s.r. spectrum for polymethylmethacrylate after exposure to γ-rays at room temperature, together with a stick diagram derived from liquid-phase data for the same radical. (Note: this scheme requires that $A(CH_3) = A(H_1) + A(H_2)$.)

Both primary and growing polymer radicals can be detected and characterised by e.s.r. spectroscopy, and rates of initiation, growth and termination can be estimated. One especially interesting intermediate is the growing polymer radical $RCH_2\dot{C}(CH_3)CO_2CH_3$ formed from methyl methacrylate. The liquid-phase spectrum for this radical gave $A(^1H, CH_3) = 22.4$, $A(^1H)CH_2 = 10.0$ and 12.7 G. Compare this result with the e.s.r. spectrum that results when polymethacrylate is exposed to ionising radiation [10.12]. This spectrum, shown in Fig. 10.1, comprises a set of 5 sharp lines, with an intermediate set of 4 lines which are apparently weaker. This spectrum is readily obtained with great intensity, even at room temperature, and hence became the centre of many early studies. Explanations ranged from pair-trapping to the presence of two species, one giving 5 lines ($A \sim 23$ G) and the other giving 4 lines ($A \sim 23$ G), both sets with binomial intensities. My own view was that a single radical species was formed, which must surely be the relatively stable growing polymer radical $RCH_2\dot{C}(CH_3)CO_2CH_3$. If the conformation of this species is one in which the polymeric R-group avoids the radical plane as in 10.21,

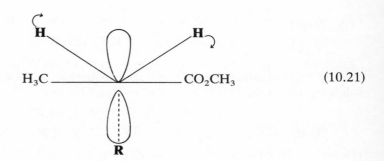

(10.21)

then, using the $\cos^2 \theta$ rule outlined in Chapter 3, the methylene protons should give a coupling of about half that displayed by the methyl protons (i.e. *ca.* 11.5 G). The resulting spectrum should thus comprise 9 lines of relative intensities $1:2:4:6:6:6:4:2:1$. This gives the correct number of lines and separations, but the intensities are incorrect. However, if the methylene protons are not quite equivalent, the intermediate lines would split into doublets. Provided the difference is small, this extra splitting would not be seen in the solid-state spectrum but alternate lines would broaden. These prognostications are clearly verified by the liquid-phase result reported above.

Although the reactions studied by Fischer [10.11] are the forward part of equation 10.18, the reverse processes constitute an important mode of polymer degradation. Often the reverse steps dominate, especially in the absence of an excess of 'monomer'. Some examples studied by e.s.r. include:

$$H_2\dot{C}CH_2S(O)R \rightarrow CH_2{=}CH_2 + R\dot{S}O \qquad (10.21)$$

and

$$H_2\dot{C}CH_2S(O)_2R \rightarrow CH_2{=}CH_2 + R\dot{S}O_2 \qquad (10.22)$$

A rather different type of β-elimination is found for radicals of type

$$\underset{\overset{|}{OH}}{\overset{}{\dot{C}}}{-}\underset{\overset{|}{X}}{\overset{|}{C}}{-} \quad ,$$

which may loose X^- and a proton to give the radical

$$\underset{O}{\overset{}{{>}\!C}}{-}\dot{C}{<} \quad ,$$

where $X^- = OH^-$, hal$^-$, $PO_4{}^{3-}$, etc. An example is the reaction between $\cdot OH$ radicals and aqueous ethylene glycol [10.13]:

$$\cdot OH + HOCH_2CH_2OH \rightarrow HO\dot{C}HCH_2OH \qquad (10.23)$$

$$HO\dot{C}HCH_2OH \rightarrow O{=}CH\dot{C}H_2 + H_2O \qquad (10.24)$$

10.2.2 α- and β-scission in phosphoranyl radicals

The reactions under consideration are summarised in the following scheme:

$$R\cdot\, + OPL_2R' \xleftarrow{\;\beta\;} \underset{\overset{|}{L}}{\overset{\overset{\displaystyle OR}{|}}{\dot{P}}}\!\!\overset{{\diagup}R'}{\underset{\diagdown L}{}} \;\rightleftharpoons\; \underset{\overset{|}{L}}{\overset{\overset{\displaystyle R'}{|}}{\dot{P}}}\!\!\overset{{\diagup}OR}{\underset{\diagdown L}{}} \xrightarrow{\;\alpha\;} R'\!\diagup\, + PL_2OR \qquad (10.25)$$

As stressed in Chapter 8, phosphoranyl radicals have been studied extensively in the liquid phase, especially by Davies, Roberts and their co-workers [10.14] and by Kochi and his co-workers [10.15]. The large coupling to ^{31}P together with the small anisotropies and consequent narrow liquid-phase features make these radicals ideal for detailed mechanistic study, the competition reactions shown in Scheme 10.25 above being illustrative of what has been accomplished. It seems that α-scission is much faster for axial R- than for equatorial R-groups, and hence that for phosphoranyl radicals in which alkyl groups prefer to be equatorial, the slow step may be 'pseudo rotation' to place R- in the unfavoured axial site from which loss of R· is facile. This follows the predicted differences in bond-strengths. In contrast, β-scission rates are if anything, marginally in favour of preferential loss from the equatorial position, although the e.s.r. evidence is less clear-cut.

10.2.3 Electron addition to phosphorus(V)

Parallel with the liquid-phase studies discussed above have been our own studies of irradiated phosphorus compounds [10.16]. Some of the results are summarised in the following scheme:

$$(10.26)$$

The choice of route seems to depend upon a range of factors, some of which are quite subtle. One factor that has not been widely explored is that of temperature: since most of these reactions are thought to be kinetically rather than thermodynamically controlled, changing the temperature could well modify the nature of the products. This is one reason why working at near absolute zero is no guarantee that those primary products of significance to room temperature reactions will be obtained. Primary radiation products may well be detected, but they often relate to quite different reactions from those that dominate at higher temperatures. In Scheme 10.26, the initial electron adduct is pictured as the parent tetrahedral molecule, since addition must precede relaxation. At this stage the location of the electron is unknown. One localising step is for a ligand to act as acceptor. This frequently occurs if one ligand is aromatic, the product being simply a substituted aromatic anion. The presence of strongly electronegative ligands such as chloride tips the scales in favour of (ii) or (iii) even in the presence of aromatic groups. Step (ii) is perhaps the most common, the process being one of bond-bending. The ligands which move are L_1 and L_2, going into the 'axial' sites. This movement is often kinetically controlled and a subsequent irreversible 'rotation' to give thermodynamically more stable species (Step (v)) may be detected on warming. Step (iii) may be

100

found if there is one distinct ligand of high electronegativity, such as —Cl or —Br. The radical is then formed by bond-stretching, the electron going into a relatively localised σ^* orbital. Step (iv) is an example of dissociative electron capture. Other reactions often observed include loss of alkyl from alkoxy ligands:

$$(RO)_3PO + e^- \rightarrow R\cdot + (RO)_2PO_2^- \tag{10.27}$$

and, occasionally, loss of Cl_2^-:

$$Cl_2PL_2 + e^- \rightarrow \cdot Cl_2^- + PL_2 \tag{10.28}$$

The range of possibilities is clearly large, and e.s.r. spectroscopy has helped enormously in sorting out these steps. Step (i) is quite definitive, of course, since the coupling to ^{31}P is very small (20–30 G). Distinction between (ii) and (iii) is not so clear, since large ^{31}P coupling constants (in the 500–1000 G range) are observed for both classes or radicals. However, the coupling to the halogen nuclei is much greater for (iii) and, most importantly, the parallel component for this coupling is coaxial with the parallel ^{31}P coupling for (iii), whereas for (ii) it lies along a perpendicular direction. It is usually easy to distinguish between these structures and the phosphoryl radicals formed in (iv). As a general rule, these give smaller ^{31}P coupling constants than those for the corresponding phosphoranyl radicals and ligand hyperfine coupling is also smaller. Generally, steps (ii) and (iv) occur simultaneously, but on annealing it is not usual for the phosphoranyl radicals formed by (ii) to be converted into phosphoryl radicals. Not enough is known about the σ^* radicals for any definitive statement, but it is most probable that Step (iii) is a precursor to (iv), as implied by Step (vi).

The discovery that electrons react either by step (ii) or (iv) of Scheme 10.26 with a range of mono-, di- and tri-alkyl phosphates [10.17], in addition to Reaction 10.27, made us wonder if such reactions could be of significance in biological phosphates. This seemed to be supported by the detection of such species from certain irradiated sugar phosphates, but AMP, ADP and ATP all failed to give evidence of electron addition at phosphorus, and we must conclude that the electron-scavenging power of the unsaturated bases is too great (*cf.* Step (i)).

10.2.4 *Dissociative electron capture*

An example of dissociative electron capture (d.e.c.) is given in Step (iv) of Scheme 10.26. This could have taken the alternative route to give

$$\cdot L_3 \; + \; :P\overset{\displaystyle L_1}{\underset{\displaystyle L_4}{\overset{\displaystyle |}{-}L_2}} \quad{}^- \tag{10.29}$$

and, indeed, this would be expected under certain conditions. In general we can write

$$AB + e^- \rightarrow \cdot A + B^- \tag{10.30}$$

$$\rightarrow A^- + \cdot B \tag{10.31}$$

as two alternative routes. In some cases, only one of these is reasonable, as in equation 10.32:

$$CH_3Br + e^- \rightarrow \cdot CH_3 + Br^- \tag{10.32}$$

In other cases the choice is far from clear and changes in the conditions can drastically change the products. One clear-cut example is for $Cl_3C\text{—}NO_2$:

$$Cl_3C\text{—}NO_2 + e^- \rightarrow \cdot CCl_3 + NO_2^- \tag{10.33}$$

$$\rightarrow CCl_3^- + \cdot NO_2 \tag{10.34}$$

Methanolic solutions at 77 K gave an e.s.r. spectrum characteristic of $\cdot CCl_3$ radicals, whilst the pure material, irradiated at the same temperature, gave only $\cdot NO_2$ radicals [10.18]. It was suggested that the stability of $\cdot NO_2$ and the large dispersal of negative charge would normally favour equation 10.34, but that solvation by hydrogen bonding would occur mainly at the oxygen atoms of the parent anion and this would stabilise NO_2^- rather than CCl_3^- (equation 10.33).

It is also interesting to compare d.e.c. with electron capture, as we have done in Scheme 10.27 for tetrahedral phosphorus compounds. Compare Reactions 10.32 and 10.35:

$$F_3C\text{—}Br + e^- \rightarrow F_3C \dot{-} Br^- \tag{10.35}$$

Methyl bromide gives only methyl radicals. Indeed in certain rigid environments a weak complex, $H_3C \cdot \ldots Br^-$, can be detected whose spectrum is clearly that of a methyl radical, but which exhibits an extremely weak coupling to ^{79}Br and ^{81}Br. This is probably largely dipolar in character [10.19] and shows that the fragments stay together within the cavity of the rigid matrix. Had an anionic intermediate represented a potential minimum, then surely, under these circumstances, this would be detected. In contrast, the e.s.r. results for F_3CBr^-, and also for the chloride and iodide, show unambiguously that these are σ^* radicals with relatively high spin-density on the halogens [10.20]. I suggest that the difference lies in the fact that $\cdot CH_3$ radicals are planar in their ground-state whereas $\cdot CF_3$ results are pyramidal. Indeed, there is probably little change in the $F\hat{C}F$ angles on going from $F_3C\text{—}hal$ to $F_3C\cdot$. Thus as the C—hal σ bond stretches to accommodate the extra electron, the CH_3 group will flatten, thus weakening the σ bond still further because the orbital on carbon changes from sp^3 to pure p. Hence no minimum is reached until the bond is effectively broken. This does not happen with the F_3C-derivatives and hence stable 3-electron bonds are formed. This also applies to a variety of other carbon-halogen bonds. For example, the anions $N\equiv C\dot{-}hal^-$, $PhC\equiv C\dot{-}hal^-$ and

are all typically σ^* radicals [10.21] and, in all cases, d.e.c. would not result in any major change in shape or orbital hybridisation at carbon. These σ^* anions

102

are, surely, the precursors to d.e.c. and certainly in the case of $F_3C\text{-}hal^-$ anions, annealing above 77 K gave clean conversion to $\cdot CF_3$ radicals [10.20].

10.3 Kinetic aspects

Two distinct methods are used to give kinetic information, the classical method of following the change in concentration of a species as a function of time, and the conversion of e.s.r. line-widths into life-times. We will not dwell on the former, although it is an extremely important method of probing the mechanisms and energetics of radical reactions in solution, since e.s.r. spectroscopy is used incidentally only and has no significance as such. Absolute concentrations are hard to obtain with any accuracy, but relative concentrations can be accurately monitored. Various rapid-reaction techniques, such as stopped-flow or, for photolyses, rotating sector methods have been used successfully. A recent review by Ingold, who is one of the leading practitioners, is listed in ref. 10.22. Many of the reactions that have been studied occur at, or close to, the diffusion controlled limit. An interesting example in which charge repulsion inhibits dimerisation is for the sequence of radicals $HO\dot{C}HCO_2H$ ($2K = 1.1 \times 10^9$ $M^{-1}s^{-1}$), $HO\dot{C}HCO_2^-$ ($2k = 2.5 \times 10^8$ $M^{-1}s^{-1}$) and $^-O\dot{C}HCO_2^-$ ($2k = 1.5 \times 10^7$ $M^{-1}s^{-1}$).

As indicated in Chapter 5, linewidths often contain significant kinetic information. Here we are only concerned with control by chemical reactions, the examples selected being typical of the systems that can be conveniently studied by linewidth measurements. In addition to the life-times of the radicals participating in the equilibria being studied, the linewidths are controlled, in the fast-exchange regime, by the separation between their absorption peaks. Thus this datum is required and hence, if possible, it is desirable to slow the reaction down sufficiently to give the slow-exchange spectra. This can be achieved by dilution or lowering the temperature.

Probably the most studied types of reaction using line-broadening are electron and proton transfer. These reactions are difficult to study by conventional methods because they are very fast, but are readily and simply studied by measuring linewidths.

Consider, as an example, electron transfer between a radical anion AH^- and the neutral molecule AH. When $[AH]$ is zero the spectrum for AH^- will comprise a narrow doublet. As $[AH]$ is increased, so the reaction:

$$\cdot AH^- + AH \rightleftharpoons AH + \cdot AH^- \tag{10.36}$$

will begin to limit the life-time of a given $\cdot AH^-$ radical and hence broadening will ensue. As discussed in Chapter 5, the broadening stems from an uncertainty in line position: if an electron jumps from an RH molecule for which $M_I(^1H) = +\frac{1}{2}$ to one having $M_I(^1H) = -\frac{1}{2}$, there is a life-time limitation. Since jumping to an RH molecule with the same proton-spin state causes no uncertainty, such jumps do not contribute. Thus in this case there is a statistical factor of 0.5. Obviously as the number of lines increases so the statistical factor will vary (i.e. $\frac{2}{3}$ for a ^{14}N triplet). Also, if lines have different intensities because of different degeneracies, the extent of broadening will fall as the degeneracy increases, because of the increased probability of that M_I quantum number being retained during electron transfer.

In reactions of this type, rates are usually measured in the region of initial broadening, in which case the appropriate equation is:

$$\Delta H = \frac{1}{2\tau\gamma e}$$

where ΔH is the width enhancement in G and 2τ is the mean life-time of the anion AH^-. (γe, the electron magnetogyric ratio, $= 1.76084 \times 10^7$ rad s^{-1} G^{-1}.) The rate, $\frac{1}{2}\tau$, relates to individual anions. If the conventional second order rate constant is required the controlling factor is the concentration of the neutral RH molecules:

$$k_2 = 1/(2\tau[\mathrm{RH}])$$

The first electron-transfer reaction studied in this way was between naphthalene and its anion [10.23]. In tetrahydrofuran at room temperature the derived rate was 5.7×10^7 l mol^{-1} s^{-1}, which is about 100 times less than the diffusion controlled limit. Many such electron-transfer processes have now been studied for anions of aromatic hydrocarbons and their derivatives, especially quinones.

In solvents of low dielectric constant the anions may be largely present as ion-pairs (Chapter 9). In such cases, hyperfine splitting to the cation nuclei is invariably retained, and is still present in the fast-exchange (exchange-narrowed) region. This means that the same cation is involved during many transfers. In other words, each electron transfer is accompanied by a cation transfer. This is frequently described as an 'atom transfer' (of M·), but in my view this is most misleading. There is no evidence that suggests that the electron-transfer path involves the momentary or even incipient formation of M· atoms, and since the rates for ion pairs are very similar to those for the anions, it is most improbable that such a major modification has occurred.

In fact, simultaneous cation transfer is a necessary requirement of the principle of microscopic reversibility, as can be seen from Fig. 10.2. If cation movement did not accompany electron movement, the process would be asymmetric, but since it is perfectly reversible this is obviously rejectable. The same considerations apply to molecular shape and to solvation.

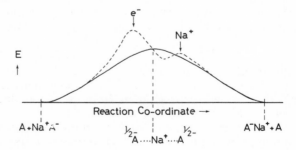

Figure 10.2 Hypothetical reaction coordinate for an electron transfer between A and the ion-pair A^-Na^+. The principle of microscopic reversibility requires that the forward path equals the reverse (as in the full curve), since exactly the same reaction is involved. This rules out an asymmetric reaction in which the electron and cation are transferred at different times (as in the dashed curve).

The 'shape barrier' is minor for anions like that of naphthalene, which probably have dimensions closely similar to those of their parent molecules. Consider, however, the transfer

$$NO_3^{2-} + NO_3^- \rightleftharpoons NO_3^- + NO_3^{2-} \qquad (10.37)$$

Here, NO_3^- is planar, but NO_3^{2-} is pyramidal (Section 7.5). Transfer to give pyramidal NO_3^- and planar NO_3^{2-} is clearly rejectable, again because of the reversibility or symmetry requirement (unless it is induced by light, in which case transfer is over before relaxation can occur, on the Franck–Condon principle). We must therefore envisage a situation in which transfer only occurs when, fortuitously, an NO_3^- ion has been induced to become momentarily pyramidal, probably by collisional activation, and an NO_3^{2-} radical is less pyramidal than usual, again as a result of some activation process. The electron will then move with no barrier once good overlap of the appropriate orbitals is attained.

Such arguments are readily extended to include solvent molecules. The anions will interact with most solvent molecules more strongly than the neutral molecules. Again, the system must wait for fortuitous equalisation of solvation energies before electron transfer can occur. Hence even fewer collisions will result in transfer. Thus strong anion solvation as, for example, of semiquinones in protic solvents, should hinder electron transfer. Strong cation solvation, on the other hand, may enhance transfer if the cations are removed, or are present as solvent-separated ion-pairs. However, for contact ion-pairs such solvation could impede transfer since partial desolvation may be needed in order to reach a symmetrical transition state. These problems have been discussed in depth by Szwarc [10.24].

Electron-transfer reactions are, of course, of fundamental importance in biology. Details of the reactions involved remain obscure in most cases, but quinone structures are often involved. Rates have to be extremely fast, and it is unlikely that considerations such as those just outlined are important. Shape changes will be small for large quinone-semiquinone systems, and probably the limitations caused by solvation are also absent or kept very low by charge delocalisation. Indeed, if the quinones are in a suitable array, the electron can move so rapidly from one to another that there will be no time for solvation at any one site to occur. This situation occurs in certain solids, being controlled by the relative rates of electron transfer and relaxation. This can again be illustrated by the reaction of NO_3^{2-} and NO_3^- (equation 10.37), this time occurring in, say, sodium nitrate crystals. Two extremes can be envisaged: in one, initial addition to a planar nitrate is followed by transfer to a neighbour so rapidly that relaxation to give pyramidal NO_3^{2-} cannot compete. In that case conductivity is high, and the description of an electron in a conduction band becomes appropriate. In the other extreme, relaxation of a given planar NO_3^{2-} is faster than transfer, in which case the localised chemical description is best, and e.s.r. spectroscopy would detect normal $\cdot NO_3^{2-}$ ions (see Chapter 7).

It is important to distinguish between these electron-transfer processes between radicals and diamagnetic molecules, and electron spin exchange between radicals. This also leads to line-broadening of hyperfine multiplets, and ultimately to exchange narrowing, and is normally avoided by using very dilute solutions ($<10^{-4}$ M). Dipolar interactions of the type discussed in Chapter 11 are also important, but provided good orbital overlap can occur in

105

collisions, electron spin-exchange normally dominates. The phenomenon is the same as before, and for a radical such as $\cdot RH^-$ the doublet features broaden, coalesce, and the singlet then narrows. It is necessary that the radicals have opposite spins for this to be effective, and as before, the nuclear orientations must differ.

The major difference between this and electron transfer is that there are now no barriers involving shape or solvation, since the exchanging radicals are identical. Diffusion control is therefore predicted, and is normally found from linewidth studies.

10.4 The photosynthetic process

In endeavouring to discover mechanistic studies involving e.s.r. spectroscopy in biochemical systems, one must turn to studies of photosynthesis. The importance of such studies cannot be overestimated since it is the Sun's radiation that, in the long run, prevents an entropically controlled descent into chaos. The photosynthetic cycle, especially that in chloroplasts, is incredibly complicated, and e.s.r. spectroscopy has only been one of many tools that have begun to establish some aspects of the whole intricate procedure (see, for example, ref. 10.25). At first sight, e.s.r. spectroscopy has not been as helpful as might have been hoped, because the major results centre around broad, structureless overlapping signals in the free-spin region. This is a false impression, as we shall see, but it has required the application of highly sophisticated methods to produce useful information.

A major problem in photosynthesis is the light induced oxidation of water:

$$2H_2O \xrightarrow{h\nu} 4H^+ + O_2 + 4e^- \tag{10.38}$$

which is coupled to the better understood reduction of carbon dioxide. It seems fairly certain that these reactions proceed by a series of electron-transfer steps, which slowly build up to the difficult task of removing electrons from water itself.

In certain bacteriochlorophyll systems water is not involved, and a relatively simple single cycle occurs, which is better understood that the double photo-system postulated for green plants. Scheme 10.39 summarises the situation as reported by Warden and Bolton [10.25]:

$$(10.39)$$

The chlorophyll molecule absorbs 870 nm light to give an excited singlet state. This may or may not become a triplet-state prior to electron transfer. (Triplet-states have been detected by e.s.r. spectroscopy, and have the relatively small D term of $0.0154 \, cm^{-1}$, implying considerable separation of the two spins.) Singlets or triplets transfer electrons to acceptor X. This is very rapid, and X must therefore be close by. It seems probable that X is a quinone molecule called ubiquinone (UQ) and that chains of many UQ molecules can transport the electron away from the parent cation. Elegant ENDOR studies have picked out the proton hyperfine couplings for the chlorophyll cation, and the results show that, in fact, a dimer cation is involved rather than a monomer. This dimerisation no doubt helps to facilitate electron transfer to UQ. UQ^- anions readily pass on their excess electrons to non-haem iron centres, whose reduced (Fe^{2+}) form has been detected at 1.4 K [10.26]. Thus cytochrome b and c have been implicated as part of the complex electron-transfer cycle.

The process for green plant chloroplasts is still more complicated, and it seems fairly certain that two distinct systems are involved, only one of which is capable of oxidising water. Suffice it to say that in System I, plastocyanin, a copper-containing protein, has been strongly implicated, Cu^{2+} e.s.r. signals being detected under certain circumstances. Also, in System II, Mn^{2+} has been detected, and is thought to play some rôle in the electron-transport mechanism.

CHAPTER 11

Polyelectron systems

Most e.s.r. studies are concerned with species having one unpaired electron per molecule, i.e. having $S = \frac{1}{2}$. Nevertheless, under favourable circumstances triplet-state species ($S = 1$) can be detected, and transition-metal complexes with $S = \frac{3}{2}$ and $\frac{5}{2}$ usually give detectable e.s.r. features (Section 12.4). Here a brief review of the spectra obtained from non-transition-metal triplet-state molecules is given. It should be stressed that for many such systems there are other techniques that may well give more significant information than e.s.r. spectroscopy which no longer has the dominating rôle that it occupies in free-radical chemistry. However, under favourable circumstances, e.s.r. spectra can be extremely informative, and provide certain data not so readily obtained by optical methods.

We must first consider the splitting between singlet- and triplet-states. For a molecule in a photo-excited state, we can depict singlet and triplet as ⑦⑪ and ⑦⑦ respectively. Generally the triplet lies below the singlet because of Hund's rule. The triplet state itself comprises three components, $M_S = \pm 1$ and $M_S = 0$, which are usually very close in energy relative to the gap ($2J$) between singlet and triplet, but which are often not exactly equal. Indeed, it is this inequality that makes their e.s.r. spectra interesting. These levels, and the effect of a magnetic field thereon, are shown in Fig. 11.1. If the singlet is above the triplet the separation $2J$, which is equal to the spin-exchange term, is positive.

It is important to distinguish between the singlet state ($S = 0$, $M_S = 0$) and the $M_S = 0$ state of the triplet level. The latter, whilst non-magnetic along z (see Fig. 11.2), nevertheless has a finite magnetisation in the xy plane.

11.1 A simple model

Let us start by considering a model in which two 'free-spin' electrons are separated by a distance R, and occupy volumes small compared with R. Their interaction is then purely dipolar, and this varies as R^{-3}. Any exchange contributions can be neglected. (This is comparable with nuclear spin-spin coupling as observed in solid-state n.m.r. spectroscopy.)

We can, in a crude fashion, consider the interaction of a second electron on the electron undergoing an e.s.r. transition. The field from this electron may be

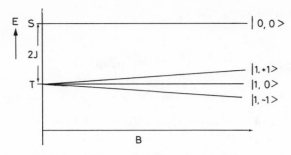

Figure 11.1 Effect of an applied magnetic field (B) on the singlet and triplet levels of a ground-state triplet molecule.

several thousand Gauss, but it falls off rapidly with separation, R. This internal field (H_i) adds to, or subtracts from, the external field and so transitions above and below the normal $S = \frac{1}{2}$ transitions are observed. Since the interaction is purely dipolar, we can write:

$$H_i = \frac{3\mu_B}{2R^3}(3\cos^2\theta - 1) \tag{11.1}$$

when μ_B is the magnetic moment of the electron. For $\theta = 0$, $H_i = 3\mu_B/R^3 = D$ and for $\theta = 90$, $H_i = -\frac{3}{2}(\mu_B/R^3) = -\frac{1}{2}D$. (This averages to zero, and so you might

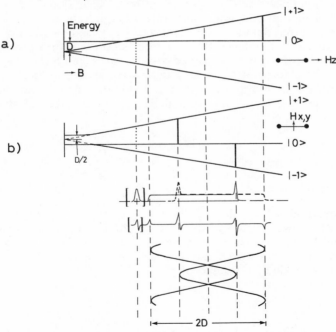

Figure 11.2 A more careful look at the triplet levels for a molecule with axial symmetry $(E = 0)$ showing the zero-field splitting D, and how this controls the form of the e.s.r. spectrum. Note how for field along x or y (perpendicular) the ± 1 levels extrapolate back to zero field to a point half way between the real zero-field levels.

SCHEME 11.1   (for $g \approx$ free spin)

$$H_z(2) = (H_0 + D')$$
$$H_y^2(2) - H_x^2(2) = 2E'(3H_0 + 2D')$$
$$H_x^2(1) - H_y^2(1) = 2E'(3H_0 - 2D')$$

(Where (1) and (2) refer to the low- and high-field x, y and z features.
D' and E' are the zero-field parameters in G.
D and E, in cm^{-1} can be calculated using $1070\ G = 0.100\ cm^{-1}$.)

$$D' = \frac{3\beta}{R^3}(3\cos^2\theta - 1)$$

for D' in G and R in Å

$$2D' = \frac{55640}{R^3}$$

(Where R is the mean separation between the two unpaired electrons.)

expect that a single central feature would appear for triplet species in the liquid phase. However, unless D is very small, the linewidth is too great and no spectrum is detected.)

Two equally intense transitions are detected, which would be coincident in the absence of dipolar coupling, but which in its presence vary with the angle between the external field and the dipolar axis (z) as shown in Fig. 11.2. This shows the angular dependence that would be observed for a properly aligned crystal containing these units together with the averaged 'power' spectrum. In Fig. 11.2(a), the external field is along the z-axis ($\parallel$), and in Fig. 11.2(b) it is $\perp$ to this (x or y). In the absence of any field, the dipolar electron-electron coupling favours electron alignment along z, and the $(0, \pm 1)$ states can be represented as

a)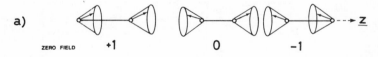

ZERO FIELD   +1   0   -1

When the external field is along z the levels change in the expected manner (Fig. 11.2(a)). However, for fields $\perp$ to z, these zero-field levels are non-magnetic and hence, as indicated in Fig. 11.2(b), they are unaffected by weak fields. However, in a strong field (H_x, say), the electrons swing round and become quantised along x instead of z:

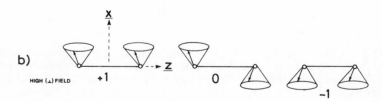

b)

HIGH ($\perp$) FIELD   +1   0   -1

If these orientations were maintained at zero field, the dashed lines in Fig. 11.2(b) would be followed. Because of the dipolar nature of the interactions, the 0 level then appears to be most stable (*cf.* the lines of magnetic field in electron-nuclear coupling for *p*-orbitals discussed in Chapter 4). Also at zero field, the extrapolated difference between the ± 1 levels and the 0 level is half that for *z* quantisation.

The link with the e.s.r. spectrum should now be clear. The simple example given here is often closely followed by radicals trapped in pairs (see Section 11.4). The zero-field parameter D (in cm^{-1}) is related to the parameter D' (in G) taken from the e.s.r. splitting which can be obtained from spectra (Fig. 11.2) by the conversion $1070 \, G = 0.100 \, cm^{-1}$.

In addition to the normal $\Delta M_S = 1$ transitions so far discussed, relatively weak '$\Delta M_S = 2$' transitions can often be detected. As can be seen from Fig. 11.2, these occur in the region of $g = 4$ (half-field) and are relatively isotropic. The intensity of these formally forbidden transitions increases rapidly as the separation between the unpaired electrons decreases. Powder spectra generally show a single feature at $g = 4$, but if the triplet can be obtained dilutely in single crystals, both this and the main doublet features may show resolved hyperfine coupling.

11.2 Non-axial symmetry

If the triplet state does not have axial symmetry, the spectrum has *x*, *y* and *z* features, as indicated in Fig. 11.3. The average coupling remains zero, and the

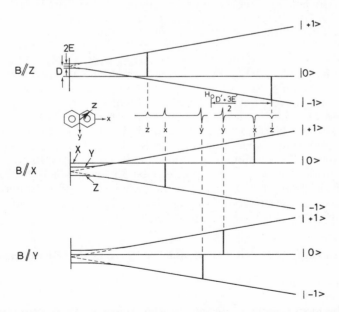

Figure 11.3 As in Fig. 11.2, but with non-axial symmetry, showing how the three levels are already split at zero field. The trends shown are close to those observed for the first triplet excited state of naphthalene.

splitting between the z features remains $2D'$. The x and y splittings are then related to E', as shown in Fig. 11.3. E' can be viewed as a measure of the deviation from axial symmetry. The zero-field splitting term, E, can again be obtained directly from E'.

An example is the naphthalene triplet-state. This has been widely studied, and was the first organic triplet-state molecule to be detected by e.s.r. spectroscopy [11.1]. If you picture this crudely in terms of each electron confined to one ring then you can see that the x, y and z coupling constants cannot be equal. The way in which field values for the detected transitions can be converted into the D' and E' parameters is indicated in Scheme 11.1. Also, a formula for calculating the mean electron-electron separation is given in this scheme.

In the absence of axial symmetry, the x, y and z components of the triplet state at zero-field (generally described as T_x, T_y and T_z) are all non-magnetic (x, y and z are the molecular symmetry axes). This is because the spin angular momenta are quenched by the molecule in a way that can be compared with the quenching of p-orbital angular momenta when an atom uses its p-orbitals for bonding. Thus the $T_{x,y,z}$ states are linear combinations of the familiar states with spin angular momentum. However, this effect is very weak, and when a magnetic field is applied along any axis, i, the T_i level moves into the $m_s = 0$ level and the other two become the $m_s = \pm 1$ levels, as shown in Fig. 11.3.

11.3 Exchange processes

When overlap between the orbitals containing the two unpaired electrons is small or zero, exchange is unimportant, and the hyperfine coupling that is detected is that of the two separate entities. For example, methyl radicals trapped say 6 Å apart in pairs (each pair being well removed from any neighbouring pairs) would give 1:3:3:1 quartets with $A(^1H) = 23$ G, just as for normal methyl radicals. However, if these could be trapped, say 3 Å apart (without recombining to give ethane), exchange would be rapid, and each electron would move effectively in the field of six equivalent protons—but would spend only half as long at each. Hence the hyperfine structure would change to a septet (1:6:15:20:15:6:1) with $A(^1H) = 11.5$ G.

Another important aspect of exchange is that if it is large, and if there is significant spin-orbit coupling in addition (i.e. if there are shifts from $g = 2.00$), then the zero-field splitting may no longer be governed solely by the spin-spin dipolar contribution. In that case the measured value of D cannot safely be used to obtain a mean separation (R) between the two unpaired electrons. (This is considered further in Chapter 12.)

11.4 Interactions between two doublet-state radicals (Di-radicals)

11.4.1 *Di-nitroxides*

Probably the most important diradicals at present are the dinitroxides, in which two $-N{\overset{\displaystyle \dot{O}}{\underset{\displaystyle R}{<}}}$ groups are held together by various chains, which can be varied so as to control, to some extent, the mean separation between the two unpaired electrons. These species generally have e.s.r. spectra that are characteristic of triplet-states. We think of di-radicals as arising when the two halves of the unit are clearly normal doublet-state radicals. Pair-trapped radicals, mentioned in Section 11.4.2 also fall into this category, but obviously photo-excited naphthalene is in no sense a diradical.

When considering 'concentrated' ($>10^{-3}$ M, say) solutions of simple nitroxides, a wide variety of types of encounter must be envisaged (Section 10.2), the measured line-broadening and ultimate production of a singlet being the net result of all such interactions. To some extent, this situation is simplified and illuminated by considering the pair-wise interactions that occur in dinitroxides. Consider the different conformations indicated in Fig. 11.4.

The e.s.r. spectrum for the fully extended conformation would comprise $\parallel$ and $\perp$ features separated by D' ($\propto R_1^{-3}$). Tumbling without conformational change would average the zero-field splitting to zero leaving, probably, a broadened triplet (from coupling to ^{14}N). As the two nitroxide groups move closer, along various (arbitrary) paths, so D increases very rapidly. Also, spin-exchange will set in as overlap becomes significant, and the 1:1:1 triplet will eventually change to a 1:2:3:2:1 quintet (Fig. 11.4).

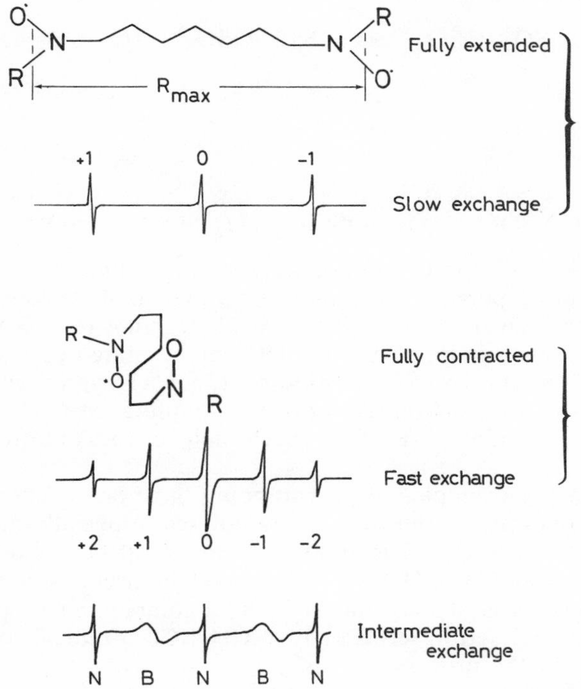

Figure 11.4 This shows how coiling of a dinitroxide radical can change the e.s.r. spectrum from that for a normal radical (triplet) to that for a diradical with fast electron exchange (quintet). If this coiling occurs at intermediate rates a marked alternation in linewidths is observed.

Since the positions of the original $M_I = \pm 1$ and 0 lines are not modified by this process, such exchange will lead to a N:B:N:B:N set of lines (alternating linewidths) as indicated since exchange is continuously varying from zero to very rapid. So modulation of the dipolar interaction leads to overall broadening, which is also controlled by Y_c, the overall rotational correlation time, whilst exchange introduces an extra pair of broad lines giving a quintet with alternating linewidths. This is indeed the sort of behaviour that is observed. The difference between this and normal exchange broadening and narrowing is that when, for example, a radical with $M_I(^{14}N) = +1$ meets another with $+1$, the $+2$ component thus formed is not shifted and hence remains narrow. Normally, however, this pair will have a short life, and both $+1$ radicals will then meet others with different values of M_I which will then result in a life-time uncertainty. Hence the narrow ± 2 and 0 lines will also broaden.

11.4.2 *Pair-trapping*

This phenomenon, usually termed 'pair-trapping', was discovered first in crystals of $K_2S_2O_8$ after ultraviolet photolysis [11.2]. After a false start, in which the radicals were thought to be $SO_4 \cdot^-$, it was realised that they were, in fact, $\dot{O}OSO_3^-$ radicals, the details of the g- and zero-field parameters being nicely accommodated by the scheme:

$$O_3^-SO—OSO_3^- \quad O_3^-SO—OSO_3^- \quad O_3^-SO—OSO_3^-$$

$$\downarrow h\nu$$

$$[SO_4 \cdot^- \quad + \quad SO_4 \cdot^-]$$

$$\downarrow$$

$$O_3^-SOO \cdot \quad O_3^-SOSO_3^- \quad O_3^-SOSO_3^- \quad \cdot OOSO_3^-$$

An initial breaking of one O—O bond gives two excited $SO_4 \cdot^-$ radicals which attack neighbouring persulphate ions to give pyrosulphate ions and the more stable peroxy radicals, $O_3SOO \cdot^-$. The crystal structure of $K_2S_2O_8$ is such that these displacements require very slight movement along one axis. The separation deduced from the zero-field splitting (15.8 Å) agrees exactly with this scheme, as do the principal directions of the coupling axes. The g-values were typical of peroxy radicals, $ROO\cdot$, which helped to confirm the proposed scheme.

Since then, many examples of pair-trapping have been reported, especially for solid-state photolyses, but also for radiolyses, especially at low temperatures. One particularly clear-cut and significant result occurs for the molecule $(CH_3)_2C(CN)N{=}NC(CN)(CH_3)_2$, which is widely used as a source of free radicals. On photolysis of the solid at 77 K, a complex e.s.r. spectrum results which was later analysed in terms of two $(CH_3)_2\dot{C}CN$ radicals trapped *ca.* 6 Å apart [11.3]. Thus the process:

$$RN = NR \xrightarrow{h\nu} R\cdot + N_2 + \cdot R$$

$$\longrightarrow \quad 6\,Å$$

occurs efficiently at 77 K.

Radiolysis gives pair-trapping far less frequently, and indeed it is perhaps surprising that it occurs at all. One example that seems to be closely similar to photolysis in that it probably stems from the decomposition of an electronically excited molecule [11.4], is:

$$\text{PhOCOOPh} \longrightarrow \text{PhO} \cdot + \text{CO} + \cdot \text{OPh}$$
$$\longrightarrow \quad 5.8 \text{ Å} \quad \longleftarrow$$

However, the primary act in γ-radiolysis is generally electron ejection, and I have suggested that pair-trapping may occasionally occur because one or more molecules adjacent to the ejecting molecule may have a very high electron-capture cross-section. This concept has been modified by Iwasaki et al. [11.5] who suggest that the ejected electron is initially captured at some distance away from the radical cation, which then transfers a proton to a neighbouring molecule. On being thermalised, the electron may return under the influence of the positive charge, but will then react with the protonated neighbour:

$$\text{SH} \rightarrow \dot{\text{S}}\text{H}^+ + e_t^-; \qquad \dot{\text{S}}\text{H}^+ + \text{SH} \rightarrow \text{S} \cdot + \text{SH}_2{}^+; \qquad \text{SH}_2{}^+ + e^- \rightarrow \text{S} \cdot + \text{H}_2$$

11.5 Ground-state triplet molecules

The most important ground-state triplet molecule is, of course, dioxygen, O_2. This has an extremely complicated gas-phase spectrum, the complexity arising from coupling between the large number of rotational levels and the spin-levels, and no attempt will be made to analyse the data here. The spectrum is described and analysed in ref. 11.6.

A very important organic intermediate having a ground triplet state is methylene, $:CH_2$. This is the parent of the carbene series, $:CR_2$, many of which have been studied in the solid state at low temperatures. These species, generally formed by photolysis of precursors such as the diazoalkanes, R_2CN_2, were originally detected by Wasserman and his co-workers, whose work has in fact dominated this field [11.7]. For methylene, the zero-field splitting (D) is extremely large (0.68 cm^{-1}), because the two electrons are confined to the one carbon atom, with little chance of delocalisation. However, the measured E term (0.0034 cm^{-1}) is far smaller than would be expected for a bent molecule with $\theta = 136°$ that has been predicted theoretically, and estimated from ultraviolet spectra. (The predicted E term is then ca. 0.05 cm^{-1}.) A hint towards the proper explanation of this surprising result is that $:CD_2$ gave $D = 0.76$ cm^{-1} and $E = 0.0046$ cm^{-1}. For stationary radicals, the D and E terms should be almost identical, but if the molecules exhibit large zero-point motions (the data were obtained at 4.2 K) a large difference is expected. Also, this motion includes a rotation about the long axis, thus explaining the small values of E. Indeed, Wasserman et al. have calculated from these differences that $\theta = 136°$. They also estimate $\theta = 137°$ from the isotropic ^{13}C hyperfine coupling. Thus the shape of $:CH_2$ seems to be settled. It is worth considering this bond angle in relation to other AH_2 species. One suitable view of the structures of these molecules is as shown in Fig. 11.5. When the in-plane sp-hybridised lone-pair orbital is empty, as for BeH_2, $\theta = 180°$. When it is half filled, as for $\cdot BH_2$ or

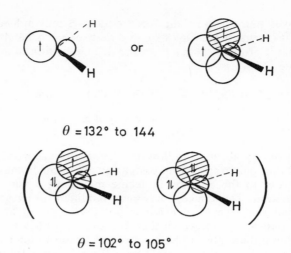

$\theta = 132°$ to 144

$\theta = 102°$ to $105°$

Figure 11.5 Shows the way in which the HX̂H angle depends on the number of electrons in the in-plane (*s-p* hybridised) orbital and not on the number in the *p* orbital normal to the molecular plane.

:CH_2, θ is between 132 and 144°, and when it is filled, as in ·NH_2 or H_2O, θ is between 102 and 105°. The number of electrons in the $p(\pi)$ orbital is not directly pertinent.

For molecules such as PhCPh, the zero-field splitting ($D = 0.4055$) is much smaller. This arises because the $2p_x$ orbital is now conjugated with the π-levels of the two benzene rings and the unpaired π-electron spends relatively little of its time close to the other electron which remains largely confined to the central carbon atom. Since the magnitude of the interaction falls off as R^{-3}, nearly all the $2D$ term arises from the π-electron when it is on this carbon atom (Fig. 11.6).

(a)

(b)

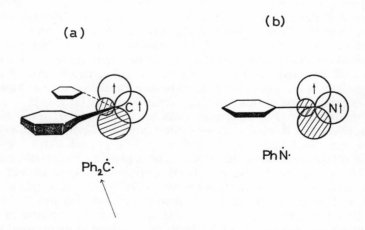

$Ph_2\dot{C}\cdot$

$Ph\dot{N}\cdot$

Figure 11.6 Structures for the ground-state triplet molecules $Ph_2\dot{C}\cdot$ and $Ph\dot{N}\cdot$.

116

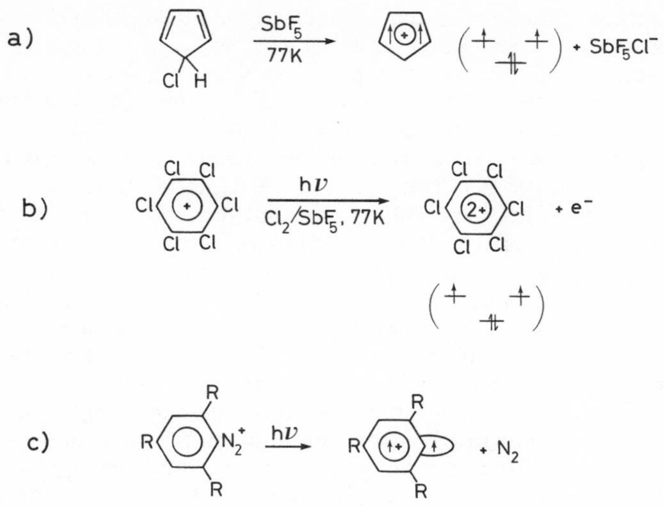

Figure 11.7 Some reactions leading to the formation of unusual organic species that exist as ground-state triplets.

Similarly for the nitrenes. For PhN· one electron is delocalised into the benzene ring whilst the other is in the in-plane $2p$ orbital and is confined to nitrogen (Fig. 11.6(b)). The data confirm this model most satisfactorily (*cf.* for PhN:, $D = 1.003$ cm^{-1}).

The two cations $C_5H_5^+$ and $C_6Cl_6^{2+}$ are of particular interest since they beautifully confirm the normal molecular orbital structures given to aromatic molecules (Fig. 11.7). In both cases, the uppermost π-levels (π_2 and π_3) are predicted to be degenerate and so the structures are $\pi_1^2 \pi_2^1 (\uparrow) \pi_3^1 (\uparrow)$. Both these four-$\pi$ electron cations have been prepared in SbF$_5$, and their e.s.r. spectra confirm that they are ground-state triplets.

For $C_5H_5^+$ $D = 0.1844$ cm^{-1} and $E = 0$. This confirms the high symmetry of the cation, showing that there is no tendency towards any distortion. The mean separation between the spins comes to *ca.* 2 Å, which fits the structure very satisfactorily. Similarly, for $C_6Cl_6^{2+}$, $D = 0.101$ cm^{-1} and $E = 0$, again confirming the high symmetry expected. The mean separation is now *ca.* 2.5 Å. Interestingly, for the triplet-state of benzene itself, $D = 0.157$. The fall on going to $C_6Cl_6^{2+}$ probably arises because of delocalisation onto the six chlorine atoms.

Another potential ground-state triplet is the phenyl cation, Ph$^+$ (Fig. 11.7). This has never been studied by e.s.r. spectroscopy, but various substituted cations, prepared by photolysis of the corresponding diazonium cations, have been shown to exist as ground-state triplets [11.8].

11.6 Photo-excited triplet-states

As I mentioned above, the pioneering work on photo-triplets was by Hutchison and his co-workers [11.1], using naphthalene dilutely incorporated in single

117

crystals of durene. Since then, many photo-triplets have been detected both by conventional e.s.r. spectroscopy and by various forms of optically detected magnetic resonance (o.d.m.r.) which prove to be far more sensitive than normal e.s.r. for the detection of photo-excited triplet-states. In these novel techniques, the phosphorescent emission spectrum from the triplet is modified by irradiating with a radio frequency or microwave field in order to induce magnetic dipole transitions between the 0, ±1 triplet-state levels. Some aspects of these optical methods have been reviewed by El-Sayed [11.9]. The field is extremely active, especially in the biochemical area. For example, optically detected zero-field resonance transitions for chlorophyll *a* and *b* have been reported [11.9], using *n*-octane as a solvent and the very low temperature of 2 K. Techniques of this type may prove to be of considerable importance in biology, by providing useful extensions of the widely used probe techniques involving conventional fluorescence measurements.

Before leaving this subject I must mention that although the majority of studies in this area have involved organic molecules, nevertheless some inorganic species have also been examined. These include the first triplet excited state of NO_2^- in sodium nitrite crystals [11.11] and the linear triplet molecules :SiNN and :SiCO prepared by reacting silicon atoms with N_2 or CO on a cold-finger [11.12].

11.7 CIDNP and CIDEP

These phenomena (chemically induced dynamic nuclear, and electron, polarisation, CIDNP and CIDEP) are considered together, although the former is an n.m.r. phenomenon. They both give mechanistic information about radical reactions, and hence relate to Chapter 10. Discussion has been deferred until now because they depend for their occurrence on pair-wise interactions of radicals to give triplet states. There are several excellent reviews of these fields, and we only have room here for the very briefest of surveys (see Section 11.13).

We need to consider the birth and death of radicals in order to see why spin-polarisation of nuclei or electrons occurs. First, however, I stress that you should not be surprised, especially with n.m.r. spectroscopy, that populations can be disturbed, since the thermal population difference is so tiny that even extremely subtle imbalances in rates can make a large difference to the observed spectra. When radicals are formed, they may come from singlet or triplet precursors. Examples are given in Fig. 11.8(a). The resulting radicals are often (but not always) formed in pairs, and these stay relatively close together for a significant time, because of the cage effect of the surrounding solvent molecules. The CIDNP and CIDEP phenomena occur in the presence of the magnetic field used for resonance studies, so we need to consider what occurs in the presence of this field (B_0). We really need an average over large and small separations of the two radicals, but it is convenient to suppose that the mean effective separation is quite large, so that the zero-field splitting of the triplet-state is relatively unimportant, and the singlet level is not far removed from the triplet manifold (J is small). This leads to the situation in Fig. 11.8(b). Transitions between S and T_0 levels are most facile because these levels are usually closest in energy, and are induced in the presence of the magnetic field

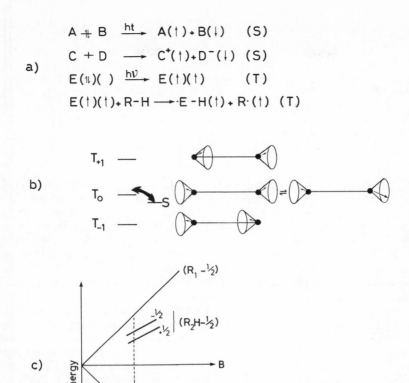

Figure 11.8 Scheme (a) gives various routes for the production of radical pairs. Thermolysis of AB and electron-transfer between C and D give singlet pairs, but photo-excitation of E followed by H-atom abstraction gives triplet pairs. In (b) the $T_0 \rightleftharpoons S$ change is indicated, and in (c) the effect of an applied field B is shown for the combined pair $R_1 + R_2H$ for the case in which $g(R_1) < g(R_2H)$ and $A_{iso}(^1H)$ is negative.

if the electron of radical 1 precesses at a different rate from that for radical 2. This will, of course, be true for radicals having different g_{av} values, but is also true in the presence of hyperfine coupling.

Consider the particular example given in Fig. 11.8(c). At a given field B_0, the biggest difference in energy (precession frequency) is between R_1 and R_2 having $M_I(^1H) = +\frac{1}{2}$. Hence when such radicals are in pairs, the $S \rightleftharpoons T_0$ inter-conversion occurs at a maximum rate, whereas the R_1, R_2 ($M_I = -\frac{1}{2}$) pairs will interconvert at a minimum rate.

We now consider ways in which diamagnetic products are formed. Often, two types are found, those from the caged radicals and those from escaped

119

radicals. An example will illustrate this point: consider the thermal decomposition of acetyl trichloroacetyl peroxide in the presence of iodine:

$$CH_3CO\!-\!OO\!-\!COCCl_3 \longrightarrow \quad \overset{\text{CAGE}}{\overset{\longleftrightarrow}{\cdot CH_3 + CO + O_2 + CO + \cdot CCl_3}} \quad \longrightarrow H_3C\!-\!CCl_3$$

$$+\ I_2 \longrightarrow H_3Cl$$

Clearly, $H_3C\!-\!CCl_3$ is a *cage product* and H_3CI is an *escape product*. Cage products can only be formed from singlet states but escape products can stem from either S or T states. Thus in our example (Fig. 11.8(c)), if we start from a singlet precursor the $R_2(+\frac{1}{2})$ radicals pass into the T_0 state faster than the $R_2(-\frac{1}{2})$ radicals. Thus the $R_2(-\frac{1}{2})$ radicals are more likely to form cage products. Since the $M_I = -\frac{1}{2}$ level is the upper level in the n.m.r. experiment, these will give an enhanced emission feature. Conversely, more $R_2(+\frac{1}{2})$ radicals will escape and so escape products will have an enhanced ground-state population and hence an enhanced n.m.r. absorption.

In our example of $CH_3CO\!-\!OO\!-\!COCCl_3$, we expect the cage product, $H_3C\!-\!CCl_3$, to be formed more for the $M_I = -\frac{1}{2}$ and $-\frac{3}{2}$ states and so should give an emission spectrum (only a single transition is observed for the methyl protons). The escape product, $H_3C\!-\!I$, should give an enhanced absorption. This is, in fact, observed. The situation would have been reversed if the proton hyperfine coupling had been positive rather than negative for $\cdot CH_3$ radicals, or if the precursor had been in a triplet rather than a singlet state. This highly simplified view seems to be qualitatively satisfactory, and has led to a set of simple rules that seem to be generally obeyed [11.13]. Quantitative estimates of the extent of polarisation are obviously extremely difficult to make because of the many subtleties involved.

CIDEP effects are less simply explained than CIDNP in a qualitative manner. It is generally agreed that there are two different, but not exclusive, mechanisms that can lead to electron-spin polarisation and hence to enhanced absorption or emission in the e.s.r. spectra of radicals. One, the 'radical-pair mechanism', is similar to that invoked to explain CIDNP effects. The other known as the 'triplet mechanism' is concerned with the way in which the radicals are initially born. The former mechanism again depends primarily on $S \rightleftharpoons T_0$ interconversion and the fact that the T_0 state cannot lead to cage reactions. The latter mechanism depends upon different populations of the triplet levels during intersystem crossing, and is especially important in photolyses [11.14]. It seems that the three states of the triplet are generated at different rates because of the rules that govern intersystem crossing. This leads to an anisotropy of spin orientation within the molecules, which is 'twisted' by the applied magnetic field into an anisotropy with respect to the laboratory-fixed axes. If separation into the two doublet states is faster than the rapid spin relaxation, this anisotropy will appear as an excess of one type of spin (α) over the other (β) in the separated radicals.

The radical-pair mechanism, which applies to many thermal reactions and probably also to most radiolyses, is closely related to the CIDNP mechanism

120

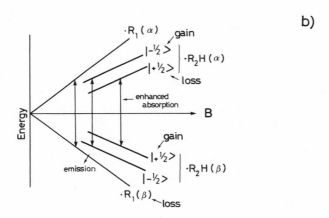

Figure 11.9 This scheme relates to the radical-pair mechanism for CIDEP. In (a) correlations between triplet and singlet levels for the radical pairs and the levels for the two doublet-state radicals are indicated qualitatively. In (b) the e.s.r. transitions for the radicals ·R_1 and ·R_2H are indicated, together with the relative gain or loss of specific proton spin states leading to enhanced absorption or emission.

described above, and can be understood in a qualitative sense in terms of the following scheme. The CIDNP mechanism alone does not lead directly to polarisation of e.s.r. spectra. However, a consideration of the correlation between the S and T states of the 'paired' or caged radicals and of the comprised states of the separated radicals, illustrated in Fig. 11.9(a) shows how this polarisation comes about. As before, the S and T states can interchange via

$S \rightleftharpoons T_0$ and since only S states can give diamagnetic cage products, there is a tendency for $[T_0] > [S]$. The energy versus field diagram in Fig. 11.9(b) then shows how the $M_I = \pm\frac{1}{2}$ levels for $\cdot R_1$ radicals and for $\cdot R_2H$ radicals diverge, given that $g(\cdot R_1) < g(\cdot R_2H)$ and that $A_{iso}(^1H)$ for $\cdot R_2H$ is negative. We conclude that $\cdot R_2H$ radicals in states $\beta|-\frac{1}{2}\rangle$ and $\alpha|+\frac{1}{2}\rangle$ are depleted relative to those in $\alpha|-\frac{1}{2}\rangle$ and $\beta|+\frac{1}{2}\rangle$ states. These will result in the low-field component appearing in emission and the high field component in enhanced absorption. (Note: the symbols α and β are used for the $\pm\frac{1}{2}$ electron-spin states to avoid confusion with the $\pm\frac{1}{2}$ proton states.)

CHAPTER 12

Transition metal complexes including those in biological systems

We can conveniently divide transition-metal complexes into those which are diamagnetic, those having one unpaired electron ($S = \frac{1}{2}$) and those having more than one unpaired electron ($S > \frac{1}{2}$). In this chapter we are concerned with the second and third categories only. Those most commonly studied by e.s.r. spectroscopy have $S = \frac{1}{2}$, since generally these give relatively well defined e.s.r. features. Then come those with $S = \frac{3}{2}$ and $S = \frac{5}{2}$, since again their features are often well defined. Complexes with $S = 1$ or 2 rarely give detectable e.s.r. features and are only briefly mentioned here.

Most courses on the chemistry of the transition metals start with an outline of the electronic states of ions and the effect of 'crystal-fields' thereon. Indeed, most e.s.r. texts discuss the spectra of transition-metal complexes on this basis, great weight being placed on the use of group theory in treating complexes of high symmetry. In addition to the definitive work of Abragam and Bleaney [12.1] this treatment is nicely developed by Wertz and Bolton [12.2]. Since this approach is not necessary for a reasonable understanding of most of the low-symmetry complexes discussed here, no attempt is made to develop this particular approach. I stress, however, that if your interest is in very ionic complexes, especially when in sites of high symmetry, a typical example being CrF_6^{3-}, then this approach is both useful and informative. It is also necessary when discussing lanthanide ions, which are not discussed here (see refs. 12.1 and 12.2.). I feel that most chemists nowadays are more interested in low-symmetry complexes in which σ and π-bonding plays an important rôle. Hence a simple molecular orbital treatment comparable with that used in Chapters 6, 7 and 8 is used. In Section 12.1 we consider the hyperfine coupling to metal nuclei and the extent to which outer s and p orbitals need to be considered together with the d-manifold. We also consider, for an unpaired electron largely confined to a single d orbital, the way in which isotropic hyperfine coupling to the metal is acquired by the 'spin polarisation' mechanism. A simple model is then described (Section 12.2) which will, I hope, give an understanding of the ways in which ligands acquire spin-density and contribute to the overall hyperfine splitting pattern. After discussing some examples with $S = \frac{1}{2}$ in Section 12.3, the extra complexities which are needed for $S > \frac{1}{2}$ are discussed in Section 12.4. In Section 12.4 a few examples are presented having

123

$S = \frac{3}{2}$ and $\frac{5}{2}$. In Sections 12.3 and 12.5 several examples are given of biological origin.

Another widely used method of probing the magnetic properties of transition metal complexes is that of measuring the magnetic susceptibility, generally by the Gouy method of weighing in an inhomogeneous magnetic field. This is in some ways complimentary to the e.s.r. method in that it gives information relating to the number of unpaired electrons in the complex, and to the orbital magnetic contributions. However, it gives no information about hyperfine interactions, which is, in my view, the most powerful aspect of the e.s.r. results.

n.m.r. spectroscopy is also used, especially to study spin-delocalisation onto ligands (*cf.* Appendix 2).

12.1 The use of metal *s* and *p* orbitals and spin polarisation

For the majority of complexes, the e.s.r. data are quite satisfactorily accommodated using the simple concept that the unpaired electrons are in pure *d* orbitals. The results in these cases establish that outer *s*-orbital population is negligible, and show that outer *p*-orbitals are only involved to the extent of a few per cent. Even when *s*-orbital participation is symmetry allowed, it is normally small, and since the outer *p*-manifold is well above the *s* level it seems that there will rarely be any need to invoke their extensive participation.

Admixture of *s* character will give a positive contribution to A_{iso}, but in fact, A_{iso} for metals is generally negative. This is explained in terms of spin polarisation (*cf.* Section 4.5). In the absence of strong covalent σ-bonding this stems from polarisation of the sets of inner *s* electrons, the overall result being equivalent to *ca.* -0.2% polarisation of the outermost filled shell (i.e. $3s$ for the first-row transition metals). Although this is a very small percentage, the actual splitting may be large because of the magnitude of A°. Some typical values for the isotropic coupling constants are given in Table 12.1.

When admixture of the outer *s*-orbital occurs ($4s$ for the first-row transition metals) the magnitude of A_{iso} falls, and the sign may even become positive, though this is rare. However, another important factor can modify A_{iso} for the metal ion which is rarely considered in the literature. This is the predominant spin polarisation mechanism discussed in Chapter 3 for π-radicals such as $\cdot CH_3$. For these species, the overriding contribution (*ca.* 4%) is from the bonding electrons. It is large essentially because the paired electrons in each C—H bond are distributed such that one is close to carbon when the other is close to hydrogen. The unpaired spin influences mainly the electrons close to carbon.

TABLE 12.1

Approximate values of A_{iso} for $3d$-transition metals in fairly 'ionic' complexes (after correcting for any orbital magnetic contributions).

Ti	^{51}V	^{53}Cr	^{55}Mn	^{57}Fe	^{59}Co	^{61}Ni	^{63}Cu
+18	−95	+20	−90	−10	−80	−33	−105

For electrons in core atomic orbitals the influence of the unpaired electron is, to a first approximation, equal for both electrons in a given core-orbital. The small (ca. -0.2%) effect that is detected relates to the extent to which the unpaired electron influences electron spin close to or away from the nucleus.

Many transition-metal complexes resemble atoms or monatomic ions, since the ligand-metal bonding is largely 'ionic' rather than being covalent. This is the case, for example, for MnF_6^{4-} or $Mn(H_2O)_6^{2+}$. However, for complexes with ligands such as —H, —CH_3, or to a less extent —NR_2, —O^- or —CN, covalent bonding is appreciable, and hence spin polarisation of the σ-bonding electrons may make a significant contribution. If, as expected, the outer s-orbital plays a part in this covalency, this effect will make a positive contribution to A_{iso} for the transition metal. The relatively small negative values for A_{iso} in many covalent complexes are probably explicable in terms of this mechanism.

For chemists interested in the e.s.r. spectra of radicals as well as transition-metal complexes, I strongly advocate adhering to A_{iso} and $2B$, used by organic and non-transition-metal inorganic chemists, for reporting data for transition metal complexes. If Δg is small, these values, divided by suitable $A°$ and $2B°$ values, give a measure of the relevant orbital populations for the unpaired electron. However, physicists use other symbols, and this can lead to some confusion. One custom sets $-\kappa = A_{iso}$, and the symbol P is related to $2B$. Others set $-P\kappa = A_{iso}$. P sometimes includes the orbital population (α^2) but sometimes this is separated out as $-P\alpha^2$, so that $-P$ relates to $2B°$. The use of these alternative symbols is described in ref. 1.

12.2 A simple model for $S = \frac{1}{2}$ complexes

We consider an 'octahedral' complex, MLX_5, having one unique, strongly bound ligand, L, and five weakly bound ligands X, and discuss the expected form of the e.s.r. spectra as the number of d-electrons is increased. Only the low-spin forms are utilised. Three properties are discussed, namely the hyperfine coupling to the transition-metal nucleus, the coupling to magnetic nuclei in ligand L and the form of the g-tensor. (If there is extensive spin-delocalisation onto a ligand, especially if it contains atoms with large spin-orbit coupling constants, this may contribute to the g-shift: this mechanism is not included in the simple model.) We distinguish three types of d-orbital having m_l quantum numbers 0, ±1 and ±2. The first, d_{z^2}, is termed the σ orbital since we are focusing attention on ligand L and its interactions. The ±1 pair comprise d_{xz} and d_{yz} and these are termed π because they overlap

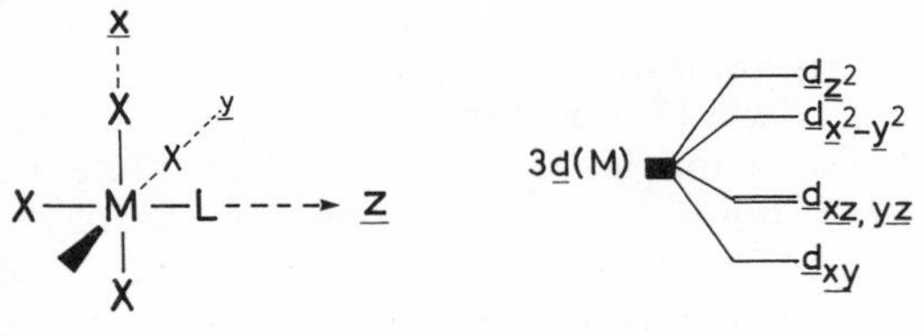

Figure 12.1 Proposed ordering of the d-orbitals in the complex MX_5L discussed in the text.

efficiently with the π-orbitals of L. The third set, d_{xy} and $d_{x^2-y^2}$ are non-bonding with respect to L, but in general will be different in energy because of bonding interactions with the four equatorial X ligands.

A probable qualitative ordering of the d-orbitals is suggested in Fig. 12.1. This is based on the assumption that σ and π-bonding to L is stronger than that to X.

12.2.1 d_{xy}^1 complexes (Fig. 12.2(a))

The anisotropic coupling to M (the metal nucleus is assumed to be magnetic) will take the usual negative sign $(-2B+B+B)$ with the z axis as parallel (Section 4.6). The isotropic coupling will probably be close to the 'normal' negative spin-polarisation value, as given in Table 12.1, unless the covalent contribution is large. Thus $A_\parallel$ and $A_\perp$ will probably be negative with $|A_\parallel| \gg |A_\perp|$. Hyperfine coupling to L will be small because of the absence of de-localisation. Probably, spin-polarisation of the ligand σ-electrons will dominate and this will be manifest as a small, negative, coupling which will usually be fairly isotropic. (This holds for ligands, such as NH_3, that bond via $s-p$ hybridized M.O.'s. For ligands such as F^-, the anisotropic coupling will normally dominate.) There is, in addition, a small dipolar coupling to L solely because of the high spin-density on M. This is a function of r^{-3} when r is the M—L bond length, and is generally close to experimental error.

The g tensor will be axial or nearly so, with its parallel component also along z. Field along z couples d_{xy}^1 with $d_{x^2-y^2}$ which will be well removed because of the five X-ligands. The antibonding $(\sigma^*)(d_{x^2-y^2})$ orbital will be the nearest and hence a small negative shift is predicted. Field along x or y couples d_{xy}^1 with d_{xz} or d_{yz} and this will usually cause a larger negative shift.

The d_{xy} orbital is π with respect to the equatorial ligands and if they have appropriate π-levels this will generally give rise to some hyperfine interaction with magnetic nuclei.

12.2.2 $d_{xy}^2 d_{xz}^1$ complexes (Fig. 12.2(b))

For our structure d_{xz} and d_{yz} are degenerate, but if the X-ligands are not identical this will not be the case. Even if they are, some distortion, to lift the degeneracy, is expected, so we will consider the unpaired electron to be confined to d_{xz}.

The anisotropic coupling to the metal nucleus will be negative, with $-2B$ along y. The isotropic coupling will probably be a relatively large negative value (from spin polarisation) and hence both $A_\parallel$ and $A_\perp$ will be negative, with $|A_\parallel| \gg |A_\perp|$. Spin density will be transferred into π_x of the ligand L, so magnetic nuclei therein will give a maximum coupling $(2B)$ along x, and A_{iso} will be small and positive from this delocalisation. However, there is an extra source of isotropic coupling to L that needs to be considered, namely from spin-polarisation of the M—L σ-electrons. This will be important if the major spin-density is on the metal, as is frequently the case. As in case 12.2.1, it will confer slight negative spin-density into the s-orbital of the L nucleus involved

126

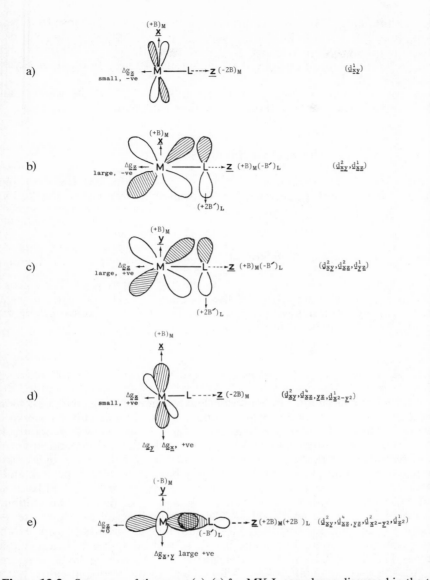

Figure 12.2 Summary of the cases (a)–(c) for MX_5L complexes discussed in the text.

and this will add to the positive contribution from spin in π_x, thus reducing the total coupling.

Another factor that needs to be remembered is that the π orbital on L may cover more than one atom, as, for example, with CN^-. In this case, both ^{13}C and ^{14}N will contribute to the spectrum. Of course, many ligands, such as NH_3, have no π orbitals, in which case the unpaired electron becomes effectively non-bonding (unless some hyperconjugation or $\sigma-\pi$ conjugation is possible).

127

The g-shift along z will now be large, since the d_{xz} orbital is close to d_{yz} and field along z couple these levels. For the d^1_{xz}, d^0_{yz} configuration this will be a shift to high-field or low g-values. The other g-shifts will be relatively small unless the orbital motion about z is strong enough to modify them, as occurs, for example, for O_2^-.

Thus, for this configuration, '$g_\parallel$' will seem to lie along z, $A_\parallel(M)$ along y and $A_\parallel(L)$ along x! However, the three interactions still share the same axes.

12.2.3 $d^2_{xy} d^2_{xz} d^1_{yz}$ complexes (Fig. 12.2(c))

The situation closely resembles that in 12.2.2. Again, we suppose some splitting of the d_{xz} and d_{yz} orbitals. The outstanding difference is that Δg_z will now be a large positive increment instead of a large negative one.

12.2.4 $d^2_{xy} d^2_{xz} d^2_{yz} d^2_{z^2-y^2}$ complexes (Fig. 12.2(e))

This final configuration is comparable with the first in many ways. The z-axis is parallel for g, $A(M)$ and $A(L)$, and the coupling to L should again be small since $d_{x^2-y^2}$ is non-bonding with respect to L. The major difference will be that $\Delta g_\parallel$ and $\Delta g_\perp$ will now be positive, with $\Delta g_\perp > \Delta g_\parallel$.

12.2.5 $d^2_{xy} d^2_{xz} d^2_{yz} d^2_{x^2-y^2} d^1_{z^2}$ complexes (Fig. 12.2(d))

The anisotropic coupling to the metal nucleus will now be positive ($+2B$ along z), and the isotropic coupling will be small in magnitude, either positive or negative, because of outer s-orbital admixture. Thus $A_\parallel$ will generally be positive and $A_\perp$ negative. Coupling to ligand nuclei (L) will depend upon the type of hybridisation in the L σ-orbital. Commonly, this is an $s-p$ hybrid so a large positive coupling to the nucleus directly bonded to M is expected, and a positive $2B'$ term along z. Thus the maximum coupling for L($A_\parallel$) will be along z and will coincide with $A_\parallel$ for the metal. Examples of ligands that exhibit a large isotropic coupling are $^{14}NH_3$, ^{13}CN and pyridine, all of which use $s-p$ hybridised σ-orbitals for bonding. Monatomic ligands such as F^- will exhibit smaller isotropic coupling constants, but these will be appreciably larger than for cases 12.2.1, 12.2.2 and 12.2.3. Nevertheless, $A_\perp$ will be small compared with $A_\parallel$. The g-component along z will be close to the free-spin value, but g_x and g_y will move to higher values because of coupling between $d^1_{z^2}$ and d^2_{xz} or d^2_{yz} orbitals.

12.3 Some examples with $S = \frac{1}{2}$

The choice of examples for this section is huge and, consequently, difficult. However, as mentioned above, a large proportion of these studies has been for strongly ionic complexes in sites of high symmetry. I have largely ignored these, since they are already well covered in standard texts, and since the more covalent, low symmetry examples are of far greater chemical and biochemical interest. I have selected several examples from biochemistry because of their intrinsic importance and because they illustrate the range of problems satisfactorily.

12.3.1 d^1 complexes

Quite the most studied d^1 complexes contain the vanadyl unit, VO^{2+}. Generally there are four equatorial ligands and sometimes a more weakly bound sixth ligand. The properties are dominated by the strong V—O σ- and π-bonds. The liquid-phase spectrum comprises an octet of lines from hyperfine coupling to ^{51}V, as illustrated in Fig. 12.3(a). On freezing solutions containing vanadyl

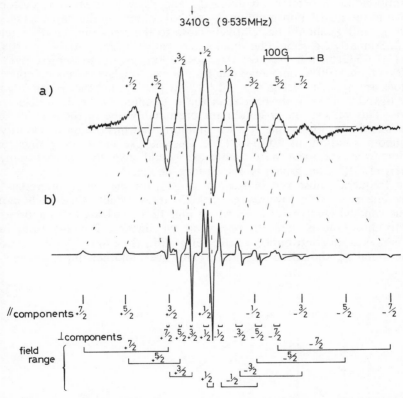

Figure 12.3 First derivative e.s.r. spectra for $VO(acac)_2^{2+}$ ions in hexamethylphosphoramide (a) just above the melting point, and (b) at 77 K. Spectrum (a) shows the isotropic spectrum with marked line-width variations stemming from the g and A-anisotropy and the relatively slow tumbling rate. Spectrum (b) shows the anisotropic spectrum and the dotted lines show how the parallel and perpendicular features average to give the isotropic lines. (Note the small splitting of the 'perpendicular' feature into x and y components.)

complexes, spectra such as that in Fig. 12.3(b) for vanadyl acetonylacetate are obtained. This shows a small splitting of some of the 'perpendicular' features into x and y components, which is expected because of the asymmetry of the equatorial ligands. The way in which the liquid-phase spectrum derives from the solid state is indicated in Fig. 12.3, and hence the curious range of linewidths in the liquid-phase spectrum can be understood. In most cases, the linewidth trends follow the solid-state spectra very well, but for the quantitative derivation of Y_c, the liquid-phase correlation time, various other factors need to be taken into account (Chapter 5). In terms of our simple model, we predict the following results for these complexes: A_{iso} should be negative, and stems from spin polarisation only. Delocalisation is likely to be small, so a 'normal' value of *ca.* -100 G can be expected. The anisotropic part $(-2B, +B, +B)$ should have z as its major axis, and if we (assume 100% spin on vanadium, we predict $A_{\parallel} = (-)\ 174$ and $A_{\perp} = (-)\ 63$ G. Because of the very strong π-bonding between the d_{xz} and d_{yz} pair and the oxygen p–π orbitals, the antibonding π-orbitals involving d_{xz} and d_{yz} will be strongly destabilised relative to the nearly non-bonding d_{xy} orbital containing the unpaired electron. Hence g_x and g_y should be relatively close to the free-spin but definitely less than 2. Similarly $\Delta g_{\parallel}$ should be small and negative. In fact, $A_{\parallel} \approx (-)\ 180$ G and $A_{\perp} \approx (-)\ 66$ G, in fair accord with expectation, giving an approximate spin-density of *ca.* 100% on vanadium for the acetonylacetonate complex. The g-values are $g_{\parallel(z)} \approx 1.945$, $g_x \approx g_y \approx 1.982$, suggesting that the π-bonding to the oxide ligand is slightly stronger than the σ-bonding to the acetonylacetonate ligands. This very strong π-bonding is presumably one of the reasons for the prevalence of the VO^{2+} unit. It is worth noting that this retains its identity even in aqueous solution, as $VO(H_2O)_5^{2+}$. This may seem puzzling since proton transfer from the axial water molecule would give the more symmetrical $V(OH)_2(H_2O)_4^{2+}$ ion, which is not, in fact, detected.

It is frequently observed that a plot of A_{iso} for the metal ion versus g_{av} is nearly linear, moving towards $g = 2.0023$ as A_{iso} falls. One such plot, for various vanadyl complexes, is shown in Fig. 12.4. At least two factors contribute to this. One is delocalisation onto the ligands. Provided spin on the ligands makes no contribution to the net g-shift, then in the limit of zero spin on the metal g_{av} should be *ca.* 2.0023. However, another important factor, in

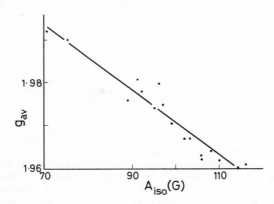

Figure 12.4 Plot of g_{av} against $A_{iso}(^{51}V)$ for various vanadyl complexes.

my view, is the degree of covalency of the M—L σ orbitals, discussed above (Section 12.1). As covalency increases, so spin polarisation of the σ bonding electrons adds a significant positive contribution to $A_{iso}(^{51}V)$. It also increases the $d_{xy} - d_{x^2-y^2}$ separation and hence reduces Δg_{av}.

Whilst on the subject of vanadyl salts, I should mention the interesting vanadyl tartrates in which two VO^{2+} units are held close together by sharing tartrate ligands. This, which gives rise to a triplet-state species of the type discussed in Section 11.4, is considered in Section 12.4.5 below.

Another group of d^1 complexes of considerable interest centre around the titanous ion having two cyclopentadienyl ligands [12.3]. Some examples are given in Fig. 12.5. These give well resolved e.s.r. spectra for both liquid and solid states, and in addition to showing satellite lines from ^{47}Ti and ^{49}Ti+cyclopentadienyl protons, extra hyperfine coupling to various other nuclei is often resolved, as indicated in Fig. 12.5. The very low symmetry for these complexes makes orbital assignments difficult, but we surmised that a $3d_{z^2}^1$ configuration was most satisfactory, since, provided A_{iso}(Ti) represents a negative coupling, $2B$ must be positive. (Since the magnetic moments for ^{47}Ti and ^{49}Ti are negative the actual isotropic coupling is really positive and $2B$ negative, but to avoid confusion I still refer to these as a negative effect and positive effect respectively.) A difficulty arises in attempts to define the z-direction. The arguments below are, however, independent of the assignment of the z-axis. Since the unpaired electron will avoid being σ- to the ligands as far as is possible, spin density on the ligands should stem from spin-polarisation of the σ-electrons. The s-character so derived for a range of ligand atoms (L), including ^{14}N, ^{31}P, ^{73}Ge, ^{119}Sn and ^{205}Pb varies between $ca.$ -0.3 and -2.8%. The isotropic coupling to ^{47}Ti and ^{49}Ti ranges from $ca.$ 6 to 11 G, decreasing as

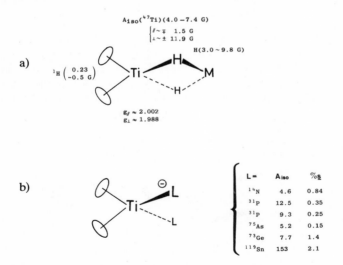

Figure 12.5 (a) Some e.s.r. results for various anionic dicyclopentadienyl titanium dihydride complexes weakly co-ordinated to their gegen ions (Na^+, Li^+, $MgBr^+$ and $AlCl_2^+$). (Note ^{47}Ti and ^{49}Ti have $I = \frac{5}{2}$ and $I = \frac{7}{2}$ and are 7.75 and 5.51% abundant respectively. Their magnetic moments are negative.) (b) Data for some $(cp)_2$ TiL_2^- complexes.

131

the spin polarisation increases. This range corresponds to an increase in coupling via the $4s$ orbital of *ca.* 3%. Since we don't know the degree of $4s$–$3d$ mixing in the σ orbitals this value cannot be used to check the spin-polarisation concept. However, if the hybridisation were d^3s this result would be reasonable.

A plot of Δg_{av} against $A_{iso}(\text{Ti})$ is linear, as for the VO^{2+} complexes. Again, this is well accommodated in terms of increased covalency leading to a fall in Δg_x and hence in g_{av}, and an increased positive ($4s$) contribution to A_{iso} from increased spin polarisation.

12.3.2 *Low-spin d^5 (ferric haem complexes)*

Our simple model suggested d_{xy}^2, d_{xz}^2, d_{yz}^1 for this case, but obviously this order depends upon the number of ligands and their symmetry. We select the important example of low-spin ferric haem complexes, typified by the partial structure shown in Fig. 12.6. Before discussing the significance of the g-values for these complexes, a few introductory comments may be helpful.

Ferric haem (d^5) systems exhibit two distinct structures, those with high spin, $S = \frac{5}{2}$, and those with low spin, $S = \frac{1}{2}$. These often occur together, in equilibrium, but the intermediate $S = \frac{3}{2}$ structure, although possible in principle, has never been detected, and is probably insignificant. The balance between these two structures is largely controlled by the sixth ligand, X in Fig. 12.6. When the Fe—X bonding is strong, the low-spin structure predominates. Thus for X = OH^-, CN^-, CO and N_3^- $S = \frac{1}{2}$ is favoured, but for X = F^- or H_2O, $S = \frac{5}{2}$

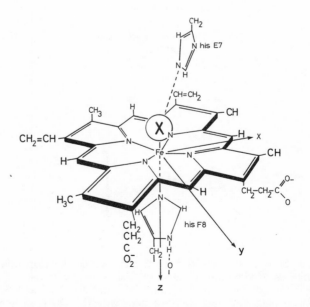

Figure 12.6 Partial structure for haem complexes in myoglobin and haemoglobin.

132

dominates. (e.s.r. spectra for the $S = \frac{5}{2}$ complexes are discussed in Section 12.4.4) The differences between these structures may well be vital to the functioning of haemoglobin. For the $S = \frac{1}{2}$ structure, in plane σ-bonding via the pyrole nitrogen atoms is strong and the iron atom is accommodated in the porphin plane. However, the in-plane ligands cannot move out radially to accommodate the extra $\sigma^*(d_{x^2-y^2})$ electron for the $S = \frac{5}{2}$ structure, so the iron atom is forced out of the haem plane. This movement, which will also involve the position of the 'proximal' histidine ligand in the fifth position, is thought to be an important lever or trigger controlling the conformation of the globin molecules.

The configuration for the low spin complexes is $d_{xy}^2 d_{xz}^2 d_{yz}^1$. The degeneracy of the d_{xz} and d_{yz} orbitals is removed by the histidine ligand, which can π-bond only with the d_{yz} level, as indicated in Fig. 12.6. This raises the π^* orbital which we write, crudely, as 'd_{yz}', above the d_{xz} orbital. Nevertheless, field along z will couple these two orbitals and hence g_z should show a large shift, as is in fact observed. Field along y couples d_{yz} with d_{xy} which is more remote than d_{xz}, so a smaller positive shift is observed. Field along x however links d_{yz} with d_{z^2}. Both the σ and σ^* orbitals involving d_{z^2} are far removed and hence g_x is close to 2.0. (If, however, g_z and g_y are large, g_x will move to low g-values—as for O_2^-, Section 2.) The results, displayed in Fig. 12.7, support these considerations. Unfortunately, it is impossible to detect hyperfine features from ^{57}Fe since the natural abundance (2.25%) is too low.

The important oxygen derivatives of haemoglobin and myoglobin are diamagnetic. However, the corresponding cobalt oxyderivatives have an extra electron, and have been extensively studied by e.s.r. spectroscopy. These are discussed in Section 12.3.4 below.

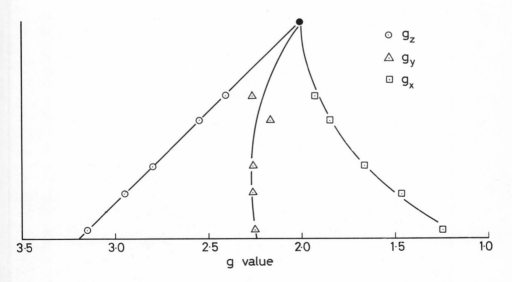

Figure 12.7 Display of g-values for a range of low spin ferric haemoproteins—$\odot g_z$, $\triangle g_y$, and $\square g_x$.

133

3.3 Low-spin d⁷ complexes

Cyanide ligands, like CO, encourage spin-pairing, and we select for discussion the two cyanides $Co(CN)_5^{3-}$ and $Fe(CN)_4NO^{2-}$. We then discuss some e.s.r. results for vitamin B_{12} derivatives.

The e.s.r. data for the two cyanides make an interesting comparison, as is shown in Fig. 12.8. The unpaired electron is unambiguously d_{z^2} rather than $d_{x^2-y^2}$, in accord with the absence of a sixth (axial) ligand. This can be judged both from the form of the g-tensor components, and of the ^{59}Co hyperfine tensor. Indeed, the qualitative predictions in Section 12.2.4 accord well with the results for these complexes. Single crystal studies show that the ligand hyperfine tensors are nearly colinear with the g-tensors, so there is no major tendency to bend, in contrast with the behaviour of the dioxygen derivatives discussed below. The ^{13}C coupling constants for the equatorial cyanides are small, showing that there is little extra delocalisation via the ring lobe of the d_{z^2} orbitals. (Note: The ^{59}Co hyperfine data had to be corrected for orbital paramagnetism before an estimate of spin-density could be made (cf. Section A2.6).)

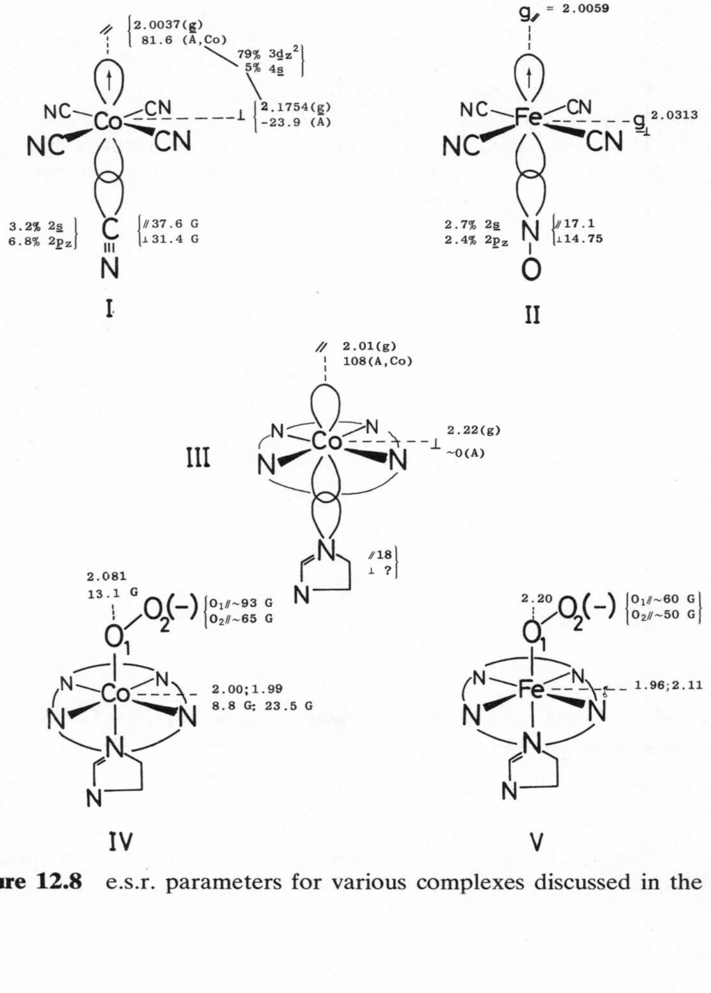

Figure 12.8 e.s.r. parameters for various complexes discussed in the text.

12.3.4 *Vitamin B_{12} derivatives*

The reduced form of vitamin B_{12}, usually described as B_{12_r} or cob(II)alamin gives a well defined e.s.r. spectrum similar to that for cobalt haemoglobin. The data are indicated in Fig. 12.8 [12.4]. The large value for $g_\perp$, with $g_\parallel$ near the free spin, together with the large $A_\parallel(^{59}\text{Co})$ and small $A_\perp(^{59}\text{Co})$ all point to a $d_{z^2}^1$ configuration. Furthermore, the presence of one coupled ^{14}N nucleus would be impossible for the alternative $d_{x^2-y^2}^1$ structure, but is expected for a $d_{z^2}^1$ electron from the coordinated imidazole nitrogen. ENDOR studies of B_{12_r} have confirmed that hyperfine coupling to the in-plane nitrogen nuclei is small, and that all proton coupling constants are also small. Vitamin B_{12_r} reacts readily and reversibly with oxygen, as does cobalt haemoglobin, to give a paramagnetic complex containing one cobalt and two inequivalent oxygen atoms. The e.s.r. results are also summarised in Fig. 12.7.

These results strongly suggest that the unpaired electron is largely confined to an oxygen π^* orbital. The g-shifts are quite reasonable for a strongly bound 'O_2^-' centre. The ^{17}O parallel coupling constants for oxycobalt haemoglobin show that the spin is not evenly distributed, but nevertheless suggest little further delocalisation. (This we can check as follows: assuming $A_\perp(^{17}\text{O}) \approx 0$ (features were not defined in the powder spectrum) we have $A_{iso}(\text{O}_{(1)}) = 31$ G, $2B(\text{O}_{(1)}) = 62$ G, $A_{iso}(\text{O}_{(2)}) = 22$ G and $2B(\text{O}_{(2)}) = 43$ G. As expected for a π^* structure, the A_{iso} values correspond to spin-polarisation only. Dividing the $2B$ values by $2B^0$ (102.2 G) we get spin-densities of 60% and 42% on $O_{(1)}$ and $O_{(2)}$ respectively.) Finally, the ^{59}Co hyperfine coupling constants are very small, corresponding to less than 4% delocalisation, and probably arise primarily from spin-polarisation effects.

All these results support the 'O_2^-' formulation. I stress, however, that this only relates to the magnetic properties—that is to say the results accord with a structure in which the unpaired electron is confined to oxygen. There may well be considerable covalent bonding involving other orbitals, and consequent charge transfer. Single crystal results show that the g- and $A(^{59}\text{Co})$-tensors have widely different directions, and support a strongly bent configuration for oxygen, as expected, although clearly the limiting end-on (η_2) configuration is not achieved.

A centre thought to be isoelectronic with cobalt oxyhaemoglobin has recently been obtained by exposing oxyhaemoglobin to ^{60}Co γ-rays at 77 K [12.5]

$$\text{HbFeO}_2 + e^- \rightarrow \text{HbFeO}_2^- \tag{12.1}$$

However, the e.s.r. data resemble those for typical low-spin ferric complexes and since, on warming above 77 K, high-spin ferric iron is eventually formed from these centres, it seems that an oxidation-reduction process has occurred:

$$\text{HbFeO}_2^- + H_2O \rightarrow \text{HbFe}^+ + OH^- + HO_2^- \tag{12.2}$$

(A proton is required for loss of peroxide, but this need not originate from a water molecule.) The g-values require a considerable spin-density on iron, in contrast with the cobalt analogue, but the detection of a parallel coupling of 50–60 G for two inequivalent ^{17}O nuclei shows that there must still be substantial spin density on oxygen.

135

12.3.5 Some d^9 complexes

By far the most important and most studied of the d^9 metal ions is Cu^{2+}. Copper(II) occurs very widely in nature and e.s.r. has been quite helpful in the task of elucidating the local structure around the cations. Silver(II) complexes are much less common, and very little work has been done on gold(II) complexes. Although in principal, there appears to be an equal probability of the unpaired electron being in $d_{x^2-y^2}$ or d_{z^2} for distorted octahedral complexes, in practice the built-in or Jahn–Teller induced distortion generally favours $d^1_{x^2-y^2}$, and $d^1_{z^2}$ configurations are rare. This is well illustrated for Ag(II) in KCl which experiences a Jahn–Teller distortion which clearly favours the $d^1_{x^2-y^2}$ structure (see below). Fortunately e.s.r. spectroscopic data show unambiguously which structure is involved.

In most cases, the results are of the type discussed in Section 12.2.5. One g-value ($g_{||}$) is found at low fields, often in the 2.1–2.3 region, the other two ($g_\perp$) being closer to the free-spin region, but still on the low-field side (*ca.* 2.01–2.03 frequently). $A_{||}(^{63}Cu$ and $^{65}Cu)$ is always $\gg A_\perp$. A typical powder spectrum is shown in Fig. 12.9, showing superhyperfine splitting from two coordinated nitrogen atoms.

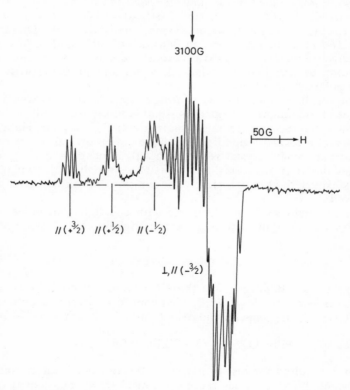

Figure 12.9 First derivative X-band e.s.r. spectrum for Cu^{2+} ions doped into $Cd[Cd(SCN)_4]$, showing hyperfine coupling to ^{63}Cu and ^{65}Cu together with coupling to two equivalent ^{14}N nuclei.

In Fig. 12.10 a range of $A_\parallel$ results are displayed as a function of $g_\parallel$ [12.10]. Only a selection of data points are included, but many other complexes fit on the main (full) lines. These lines have been drawn for $A_\parallel = 290$ and 260 G, and $2B = -170$ G, for the hypothetical $g_\parallel = g_\perp = 2$ situation, and incorporate the orbital magnetic contribution to $A_\parallel$ that is induced as $\Delta g_\parallel$ increases (Section A2.6), assuming that $\Delta g_\perp$ remains small. Clearly, this means that in the absence of a large $\Delta g_\parallel$, the hyperfine data for all these complexes would be remarkably constant, indicating that the degree of σ delocalisation remains small. Thus for the wide range of complexes close to these lines, it is $\Delta g_\parallel$ that is the significant variable, not $A_\parallel$. The $2B$ value of -170 G corresponds to a d orbital population of ca. 0.75 for these complexes.

Deviations from this major line arise for a variety of reasons, and it is noteworthy that a second set of results cluster around the lower (dashed) line in Fig. 12.10. All the complexes falling close to this line can be considered to be distorted tetrahedral species rather than octahedral or square-planar. Extrapolation to $g_\parallel = 2.00$ gives $A_{iso} = -37.4$ and $2B = -152.6$ G for these 'tetrahedral' species. Clearly the orbital of the unpaired electron is still $d_{x^2-y^2}$ primarily, so that the distortions are all towards the square-planar structure. The value for $2B$ is quite normal for Cu(II) complexes (giving ca. 67% d-orbital spin density), but the A_{iso} value is very small. We have suggested that this occurs because of direct $4s$ orbital involvement. This makes a positive contribution to the 'normal' value of ca. -100 G, an admixture of ca. 4% being required. Such admixture is forbidden in the square-planar or octahedral configuration but becomes allowed in the low-symmetry states envisaged. It should be noted that when Cu^{2+} is trapped in a fixed tetrahedral geometry, as in zinc oxide crystals, the e.s.r. properties are quite different, and spectra can only be detected at ca. 4 K or below. Presumably Jahn–Teller distortions are small because of the rigid structure of the host, and so orbital degeneracy is nearly preserved.

Cu(II) complexes differ from most others in that the normal trend in the A_{iso} vs g_{av} plots is reversed. As shown in Fig. 12.4, A_{iso} usually decreases in

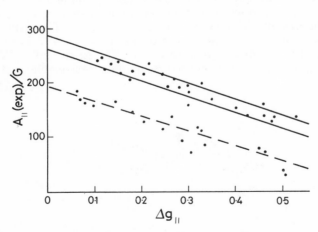

Figure 12.10 Plot of $\triangle g_\parallel$ against $A_\parallel$ (experimental) for a range of Cu^{2+} complexes discussed in the text.

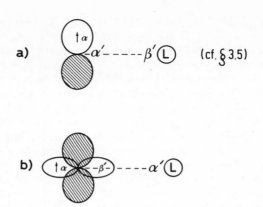

a) (cf. § 3.5)

b)

Figure 12.11 (a) Normal (π) spin-polarisation effect giving positive spin-density for the σ-electrons close to the metal and negative spin-density close to the ligand. (b) Spin polarisation induced by a σ^* electron, giving negative induced spin in the σ electrons close to the metal, and positive spin-density close to the ligands.

magnitude as g_{av} moves towards free-spin. However, for most Cu(II) complexes A_{iso} increases in magnitude. This result would be difficult to understand in terms of the usual arguments invoked to explain correlations cuch as that in Fig. 12.4, but has a ready explanation in terms of the spin-polarisation concept given above, which includes a contribution from the valence electrons of covalently bonded ligands. This is because the unpaired electron is now σ rather than π and so the sign of the effect is reversed. (Compare (a) and (b) in Fig. 12.11.) This result gives strong support to the theory that valence electron spin polarisation is indeed a significant factor, and hence that changes in A_{iso} provide a measure of changes in ligand covalency.

Hyperfine coupling to the 'in-plane' ligand nuclei fit in well with this picture. For example, the four coordinated nitrogen atoms for copper phthalocyanine [12.11] are equivalent, and show their maximum coupling along one or other of the x, y ($\perp$) directions, and their minimum coupling along the parallel direction. The results are summarised in Fig. 12.12. The estimated σ-spin density on each of the four nitrogen atoms is then ca. 3% $2s + 6.1\%$ $2p$, i.e. 36.4% total. The estimated $p:s$ ratio for ^{14}N is ca. 2.03. Whilst this is admittedly approximate, it is nevertheless quite reasonable and encourages

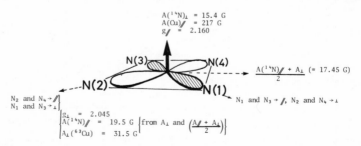

Figure 12.12 Summary of data for copper phthalocyanine showing the way the parallel direction for copper is 'perpendicular' for nitrogen.

138

faith in the method. (*Note:* the spin density on Cu, deduced from the hyperfine coupling is *ca.* 70%, which predicts a ligand density of *ca.* 30%, in fair agreement.)

Silver(II) ions give comparable results, except that the g-shifts are usually considerably greater because of the increased spin-orbit coupling constant. An interesting example, mentioned above, is for Ag^{2+} in KCl crystals [12.12]. When these centres are prepared from KCl doped with Ag^+ ions, by exposure to ionising radiation, they are formed (presumably) in an initial environment that is truly octahedral. Axial distortion is then induced, which lifts the degeneracy of the d_{z^2} and $d_{x^2-y^2}$ orbitals, favouring the configuration $d_{z^2}^2$, $d_{x^2-y^2}^1$. The resulting e.s.r. spectra, at low temperatures, show the typical axial spectrum for this configuration, with superhyperfine splitting from four strongly coupled and two weakly coupled chlorine nuclei, as expected.

The important result is that on warming, after line broadening, the spectrum becomes isotropic. This can be understood in terms of a rapid vibrational motion that moves the axial distortion between the x, y and z ligands. The orbitals follow these distortions and hence the unpaired electron becomes effectively isotropic. This constitutes an excellent example of the dynamic Jahn–Teller effect.

12.3.6 *Some Cu(II) proteins and related systems*

Copper(II) occurs widely in nature and is usually part of complex protein systems most of which have not had their structures elucidated by X-ray diffraction techniques. Many of these are intensely blue, such as ceruloplasmin, the laccases and ascorbate oxidase. These frequently have e.s.r. spectra that reveal the presence of two different types of Cu(II) centres. Type 1 centres have an unusually small value for $A_\parallel$ (^{63}Cu), which fit onto the lower, dashed line in Fig. 12.10, thus strongly suggesting a distorted tetrahedral structure that seems to be favoured by most workers. It is this centre that has a high intensity band in the 600 nm region that confers the blue colour on these proteins. Type 2 copper centres have $A_\parallel$ and $g_\parallel$ values that fall close to the two full lines in Fig. 12.10, and thus are thought to have distorted-octahedral or square-planar configuration. These centres occur together in, for example, *Polyporus versicolor* laccase, *Chenopodium album* plastocyanin and ascorbate oxidase. Others such as human ceruloplasmin only give type I spectra, and yet others display only type II spectra, examples being superoxide dismutase and galactose oxidase. In some cases the copper centres are fairly close together in the protein, in which case the e.s.r. spectra may be drastically modified. This situation is discussed briefly below (Section 12.4.5).

12.4 Complexes with $S > \frac{1}{2}$

This section is, in a sense, an extension of Chapter 11. We consider, briefly, $S = 1$, $S = \frac{3}{2}$ and $S = \frac{5}{2}$ systems, omitting the $S = 2$ case since e.s.r. signals from such complexes are extremely rare. The $S = \frac{5}{2}$ examples are the most common and most important, and there are several important biological examples including methaemoglobin, which is discussed in Section 12.4.4.

139

12.4.1 $S = 1$ complexes

There are two major differences between the $S = 1$ systems discussed herein and those usually encountered in transition-metal complexes. One is that the two electrons are spatially well separated for many triplet-states, but are more or less confined to the metal ion in the complexes. The other is that interaction is no longer largely a spin-spin phenomenon, since orbital angular momentum induced by spin-orbit coupling can provide a powerful mechanism for magnetic coupling between the electrons. These two factors often result in a complex having no detectable e.s.r. transition (especially at X-band frequencies), since the zero-field splittings are too great.

However, just because the electrons are largely confined to the metal ion, it is quite possible for the symmetry to be such that the net interaction is close to zero. It is the ligands that are responsible for constraining the electrons into different spatial regions, and high symmetry gives a spherical distribution with no zero field splitting. Thus, for a tetrahedral d^2 complex, the electrons occupy the $d_{z^2}(\pi)$ and $d_{x^2-y^2}(\pi)$ orbitals with a net dipolar interaction that averages to zero. Small distortions from true tetrahedral symmetry will result in small fine-structure splitting, that is in small values for D and E (see Chapter 11). When the symmetry is low the electron distribution is polarised and the spin-spin interaction will differ along the x, y and z axes.

Good examples of this situation are the d^2 ions, FeO_4^{2-} and MnO_4^{3-} [12.13]. These have been studied in chromate and vanadate host lattices and they probably take on the distortions of the host anions, which are dictated by the way the ions are packed in the crystal, and the resulting fields from the cations. Both ions gave the expected doublets which established conclusively the presence of two unpaired electrons, the zero-field constants being more a property of the host crystals than of the ions themselves. The g-values were close to 2.00, as expected, and the ^{55}Mn hyperfine coupling was nearly isotropic, reflecting the ^3S state of the MnO_4^{3-} ion.

These triplet-state complexes are the exception rather than the rule, and in many cases X-band e.s.r. spectra cannot be obtained. We now turn to $S = \frac{3}{2}$ and $S = \frac{5}{2}$ complexes, which nearly always give detectable signals.

12.4.2 $S = \frac{3}{2}$ complexes

In discussing results for these complexes, and in Section 12.5.3 for $S = \frac{5}{2}$ complexes, it is convenient to distinguish between two extremes, namely those with small zero-field splitting ($D < h\nu$) and those with large zero-field splitting ($D > h\nu$). These two cases are often fairly simple, but complicated spectra can arise in the intermediate regions. It is frequently helpful to measure the spectra at X- and Q-band frequencies especially for powder spectra, since they can change markedly and these changes help to identify the features. One of the difficulties is that relative intensities can vary widely, in contrast with most e.s.r. transitions. The diagram shown in Fig. 12.13 helps to predict the general form of the spectra, and we confine ourselves to a brief consideration of its implications. The reader is referred to other texts for a full exposition of the complex underlying theory [12.1,12.2].

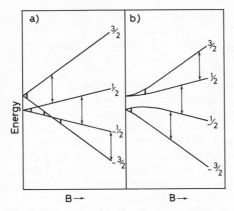

Figure 12.13 Way in which the energy levels of a complex with $S = \frac{3}{2}$ and $E = 0$ diverge with increasing magnetic field, B. (a) $B\|z$ and (b) $B\|x$ or y. The short thick vertical lines represent transitions if $D > h\nu$ and the long thin lines are for $D < h\nu$.

The first diagram (12.13(a)) displays the way the levels are split in zero-field and the influence of field along the parallel (z) axis. It is the analogue of the triplet-state diagrams given in Chapter 11, and is self explanatory. However, as the field is imposed along x or y the lower pair of levels that became $\pm\frac{1}{2}$ for $H_\|$ behave initially as if they were $\pm\frac{3}{2}$ levels, diverting rapidly as in Fig. 12.13(b). The upper pair are initially non-magnetic but bend upwards to give the $+\frac{3}{2}$ and $+\frac{1}{2}$ levels as the spins become aligned along the field instead of along the

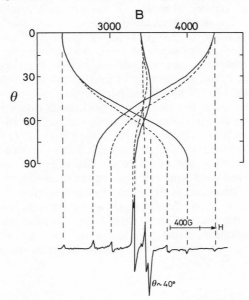

Figure 12.14 Link between the e.s.r. powder spectrum (first derivative) for Cr^{3+} (doped into $(NH_4)_2$ [$In(H_2O)Cl_5$]) and the angular variation of the resonance on going from parallel to perpendicular. Note the splitting of the central strong features. (Taken with modifications from ref. 12.14.)

141

molecular axes. The short thick vertical lines indicate the allowed transitions in the limiting case for $D < h\nu$ and the long thin lines are for $D > h\nu$. This shows that when the zero-field splitting is small, three strong transitions are observed. (This is easily remembered because the number of transitions equals the number of unpaired electrons.) However, when the zero-field splitting is large, only one allowed transition is detected having 'g' values between 2 and 4, with '$g_{\parallel}$' = 2 and '$g_{\perp}$' = 4. (The other transitions are usually at fields beyond the reach of normal spectrometers.) Both these limiting situations are often found in practice. If the ground-state is symmetrical ($^4A_{2g}$), then D and E terms will generally be small and three transitions centred on $g = 2$ are detected. (For Cr^{3+} in MgO the spectrum is completely isotropic.) However, complexes with large built-in distortions, the limiting effective g-values of 2 and 4 are observed. A nice example of the former is for Cr(III)(d^3) doped into $(NH_4)_2[In-(H_2O)Cl_5]$ crystals. A typical powder spectrum is shown in Fig. 12.14 (^{53}Cr hyperfine components are not shown in this spectrum). The way in which the crystal data link to the powder spectrum is indicated in this figure. Note, in particular, the splitting of the strong central ($\pm\frac{1}{2}$) feature into a doublet. This arises because of the turning point at ca. 40° that is in addition to the two 'normal' extrema at 0 and 90°, and is characteristic of relatively small D and E terms. From these data, D and E were estimated to be ca. 640 G and 50 G respectively [12.14].

12.4.3 $S = \frac{5}{2}$ complexes

The situation, though more complex, nevertheless resembles that for $S = \frac{3}{2}$. Since each d orbital is half populated, the ion must be ^6S, and hence the g-values are usually close to 2.0 and the A-values are nearly isotropic. For high symmetry, D and E are small or zero. Any asymmetry, however, results in a zero-field splitting that can become very large. The splitting is now into three sub-states that relate to the $\pm\frac{1}{2}$, $\pm\frac{3}{2}$ and $\pm\frac{5}{2}$ states in high-field. Thus, for an axial distortion and for H along the parallel (z) axis, the situation shown in Fig. 12.15 may be realised. For this orientation, small distortions will result in a 5-line spectrum, and as usual the outer lines will reflect the anisotropy of D, but not the central ($\pm\frac{1}{2}$) line. However, for large values of D a limiting situation is reached, just as for the $S = \frac{3}{2}$ case, but this time the value for the effective '$g_{\perp}$' feature is close to 6 rather than 4. The effective $g_{\parallel}$ remains ca. 2.0. Extremely complicated situations can arise for intermediate values of D and E.

Another limiting situation is frequently observed for $S = \frac{5}{2}$ complexes, in which a fairly isotropic single line appears at ca $g = 4.3$. It seems [12.16] that this occurs when λ, the ratio of E to D, is close to $\frac{1}{3}$, and that just this limiting situation is quite common, since it is a limiting case for rhombic symmetry. Obviously, care must be taken in distinguishing between this situation for $S = \frac{5}{2}$, and the $g_{\perp} = 4$ limit for $S = \frac{3}{2}$ [12.15].

12.4.4 Met-haemoglobin and related species

Myoglobin and haemoglobin in the oxidised iron(III) state can either be in the high-spin ($S = \frac{5}{2}$) or low-spin ($S = \frac{1}{2}$) states depending upon the nature of the

142

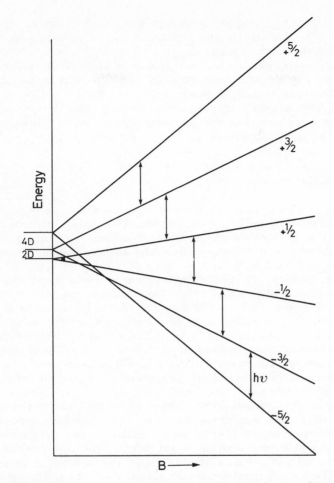

Figure 12.15 Divergence of levels with magnetic field for an $S = \frac{5}{2}$ complex having $E = 0$, and B parallel to the symmetry axis, z. The short, thick line is for $D > h\nu$ and the long, thin lines are for $D < h\nu$.

ligand in the sixth position. Indeed, the balance is so subtle that both forms frequently co-exist, although the intermediate $S = \frac{3}{2}$ state has never been detected. Examples having $S = \frac{1}{2}$ were discussed in Section 12.3.3. The natural aquated forms, generally called met-haemoglobin or met-myoglobin, exist predominantly in the high-spin form, and the e.s.r. spectra show at once that there is near axial symmetry, with $D > h\nu$, since the limiting $g_\parallel(\text{eff}) \approx 2$ and $g_\perp(\text{eff}) \approx 6$ has been reached. Single crystal studies of this signal in met-haemoglobin were of great importance in the very difficult task of solving the molecular structure of this protein, since the haem plane axes were very accurately located by their e.s.r. spectra [12.16]. The results show that $D = 9.26 \text{ cm}^{-1}$ for met-myoglobin. Also, there is a small E term ($ca. 0.02 \text{ cm}^{-1}$) which probably arises because of the asymmetry of the histidine ligand (Fig. 12.6).

The e.s.r. spectral features are usually too broad to reveal any hyperfine coupling to ^{14}N, although this is expected to be quite large because of the unpaired electrons in the d_{z^2} and $d_{x^2-y^2}$ σ^* orbitals. However, Scholes [12.17] managed to grow single crystals of perylene (a large planar aromatic hydrocarbon) containing ferriprotoporphyrin IX dimethyl ester as the chloride and hydroxide derivatives, and at 4.2 K the e.s.r. features showed the 9-line hyperfine pattern expected from coupling to four equivalent nitrogen nuclei. The coupling was almost isotropic and indicated a $2s$-spin density of ca. 2.7%. This confirms the dominance of the $d_{x^2-y^2}$ electron.

12.4.5 Metal ion clusters

This section is an extension of Section 11.3, in which pairwise interactions between doublet-state radicals were considered. Pairwise interactions between paramagnetic metal ions are frequently encountered, two important situations being the sharing of bidentate ligands and relatively high doping levels in diamagnetic host crystals.

The former type of pair-interaction has been reviewed by Smith and Pilbrow [12.18]. The principles involved are the same as those for pair-trapped radicals (Section 11.3), but we have to consider the effect of g-value variation on the spectra, and on the magnitude of the zero-field splitting. The former problem is relatively trivial and the diagram in Fig. 12.16 serves to illustrate and establish the way in which spectra are modified. Such spectra are obviously more difficult to identify than the symmetrical spectra obtained from radical pairs (cf. Fig. 11.2), but once it is realised that they are due to pairs of ions, the interpretation is straightforward.

It is important to consider different types of interaction and the way this can control the sign and magnitude of J, the singlet–triplet separation. Two extremes can be envisaged. In one, overlap between the separate orbitals of the two unpaired electrons is zero, as in (a) in Fig. 12.17. The exchange term is small or zero, but using Hund's rule, we expect the triplet level to lie below the singlet, and J is then defined as negative. However, when overlap is possible, as in (b), provided a strong bond is not formed, the singlet level may be more, or less, stable than the triplet, and as the ions approach, the singlet will move below the triplet. (Under these conditions, cooling will increasingly populate the singlet, and the e.s.r. signal intensity will fall.) Generally ligands are involved in holding the ions close together, and will frequently also be involved in governing the effective overlap and hence the sign and magnitude of J. When J is positive, the interaction is sometimes described as 'ferromagnetic'

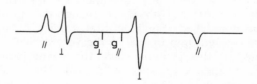

Figure 12.16 Typical first derivative e.s.r. spectrum for a pair of weakly interacting $S = \frac{1}{2}$ complexes having $I = 0$ and $g_\perp > g_\parallel$.

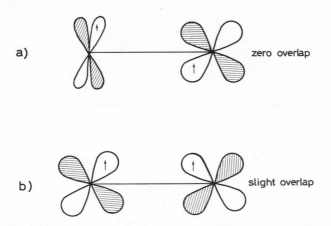

Figure 12.17 Possible modes of interaction between two doublet-state complexes when in (a) orbital overlap is zero, and in (b) orbital overlap is significant.

and when J is negative, as 'antiferromagnetic', though these terms really only apply to interactions of many ions rather than to isolated pairs.

Another problem that arises is the way in which pairwise interaction can modify the g- and A-tensors of the parent complexes. When interaction is weak, as in (a), exchange is slow, and each electron may be thought of as being confined to one of the components of the pair. Under these circumstances the hyperfine coupling and g-values will be almost identical with those of the individual complexes.

As exchange increases, the electrons become delocalised, and the hyperfine coupling to the metal nuclei changes to the pattern expected for two equivalent nuclei (i.e. two Mn nuclei would give 11 lines) and the hyperfine splitting falls to half the original value. Nevertheless, the g-values, and the principle directions of the g- and A-tensors may not be modified. If the ions are close together, and overlap between orbitals not initially containing the unpaired electrons becomes significant, there may well be a switch in the ordering of the levels, the unpaired electrons moving into the σ and σ^* levels thus produced. If this occurs, then of course the form of the g- and A-tensors can change drastically. So far as I am aware, such a situation has not yet been observed experimentally.

An interesting example of such pairwise interaction is for aqueous solutions of vanadyl tartrate [12.19]. E.s.r. spectra for vanadyl (VO^{2+}) complexes were discussed in Section 12.3.1, above. In the tartrates, two vanadyl ions can be quite rigidly held together in various ways, depending upon the pH of the system, and hence well defined triplet e.s.r. spectra can be obtained. Two such structures have relatively small values of J, which are sufficiently isotropic and sufficiently constant under the buffeting and tumbling of the liquid phase, that weak satellite features from the normally forbidden triplet-singlet transitions have been observed. These transitions are depicted in Fig. 12.18 together with a typical spectrum, showing doublet features for two different complexes (one separated by *ca.* 1400 G and the other by *ca.* 2400 G). The central region is complex because there are at least two sets of lines present, and because the

145

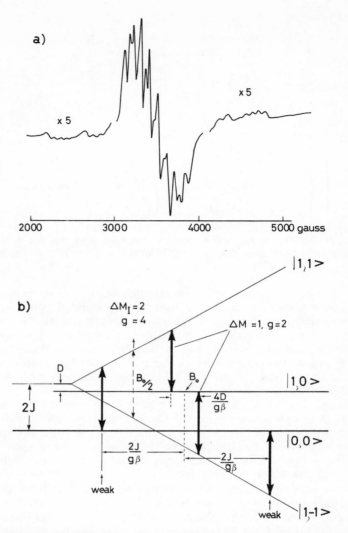

Figure 12.18 (a) First derivative e.s.r. spectrum for an aqueous solution of vanadyl tartrate at pH 3.0, showing the weak satellite lines caused by singlet-triplet transitions shown in (b).

complex is so large that rapid averaging of the anisotropic components is not quite complete. Also, although the features approach the fast exchange pattern of 15 lines, J is probably not yet large enough to give the simple uniform pattern for fast exchange.

The hyperfine patterns of the satellite features, which are *ca.* $\frac{1}{50}$th of the intensity of the central lines, are extremely complex, but were quite well reproduced by the application of simple theory [12.19]. This led to J values of 680 G and 1190 G for the two complexes. Observation of these triplet-singlet satellite features is most unusual. They were observed in this case partly

146

$$D_{max} = 180 \times 10^{-4}\ cm^{-1}$$

$$(Et_2NCS_2)_2Cu \underline{\hspace{2cm}} Cu(S_2CNEt_2)_2$$

$$(Et_2NCS_2)_2Ag \underline{\hspace{2cm}} Ag(S_2CNEt_2)_2$$

$$D_{max} = 683 \times 10^{-4}\ cm^{-1}$$

Figure 12.19 Zero-field splitting constants for dimeric *bis*-(diethyldithiocarbamato), copper(III) and silver(II).

because of the large overall hyperfine splitting for vanadium, and partly because of the great rigidity of the dimer complex.

We see that pairwise interactions, although complex, can often be understood in considerable detail. The situation becomes far more complex if the two metal ions are not the same, and if they have $S > \frac{1}{2}$. Even more complicated situations arise for clusters containing more than two such ions, and the reader is referred to the literature [12.18] for further details. Before turning attention briefly to biological systems in which several metal ions are implicated, we should stress one factor that is sometimes overlooked.

For pair interactions between radicals with g-values close to 2.0, the zero-field splitting is controlled solely by the spin–spin dipolar coupling, and is independent of the value of J. Hence good values for the mean separation between the two electrons can be calculated. However, if the g-values deviate from 2.0, as is of course frequently the case for transition-metal complexes, then an extra contribution to the zero-field splitting comes in, which can either increase or decrease the experimental values. This is well illustrated by the data for the two complexes shown in Fig. 12.19 [12.20]. Although the metal-metal separation must be about the same in each case, the values for D_{max} ($+180 \times 10^{-4}\ cm^{-1}$ and $+683 \times 10^{-4}\ cm^{-1}$ respectively) differ very greatly. This was interpreted in terms of the far greater value of the spin-orbit coupling for silver which makes the spin-orbit contribution to D very much greater for this complex.

12.4.6 *Metal ion clusters in biological systems*

In view of the need to transport electrons rapidly, often over considerable distances, and in view of the need for great selectivity in biological reactions, it is not surprising to find that groups of metal ions can often do the job better than a single metal complex. In some cases, it seems that two ions are sufficient, but in others four or more seem to be implicated.

One example involving pair interactions is found in haemocyanin, which is the oxygen carrier in the hemolymph of **arthropods** and **molluscs**. This binds one oxygen molecule per two copper atoms, and the following reaction has been suggested to explain the strong binding:

$$Cu^+ \ldots Cu^+ + O_2 \rightleftharpoons Cu^{2+}-O^{2-}-O-Cu^{2+} \tag{12.3}$$

This is supported by the blue colour of the oxy-derivative and by the oxygen-oxygen stretching frequency which is close to that expected for an O_2^{2-} ligand.

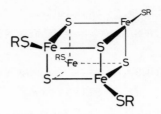

Figure 12.20 Possible structure for the $[Fe_4S_4(SR)_4]$ tetramer cluster thought to be important in iron-sulphur proteins.

However, no e.s.r. spectrum is obtained either before or after oxygenation. This suggests that the singlet state is far below the triplet state for the Cu^{2+}—O^{2-}—O—Cu^{2+} complex. This result is important since it shows that great caution is needed in assigning valence states to complexes using e.s.r. spectroscopy as a criterion, especially when weak metal-metal interactions are involved. It is interesting that when an extra electron is added to the system (on exposure to ^{60}Co γ-rays) [12.21], one Cu^{2+} ion is reduced to Cu^+, and the other gives rise to a well defined e.s.r. signal which can be used to obtain structural details about the complex.

Of wider importance are the iron-sulphur proteins such as the ferredoxins. These act as electron carriers in a range of metabolic reactions. These are thought to have monomer $[Fe(SR)_4]$, dimer $[Fe_2S_2(SR)_4]$ and tetramer $[Fe_4S_4(SR)_4]$ structures, the last of these having T_d or D_{2d} symmetry, with each iron atom bonded to three bridging sulphur atoms and one thiolate (cysteine) ligand, as in Fig. 12.20 [12.22]. These interesting structures have been nicely copied using synthetic analogues, and can exist in three different oxidation states, namely $[Fe_4S_4(SR)_4]^{3-}$, $[Fe_4S_4(SR)_4]^{2-}$ and $[Fe_4S_4(SR)_4]^-$. For ferredoxin these are written Fd_{red}, Fd_{ox} and Fd_{s-ox} respectively. Fd_{ox} gives no e.s.r. spectrum, even at 4.2 K, and Mössbauer spectra suggest that despite the fact that each iron is high-spin Fe(III), the electrons couple to give an $S = 0$ ground-state. The addition of one electron to give Fd_{red} then gives $S = \frac{1}{2}$ and an e.s.r. signal is obtained, with $g = 1.944$. This isotropic signal confirms the high symmetry and the postulated delocalisation of the effective spin. A cluster orbital theory has been developed which explains the Mössbauer and e.s.r. results satisfactorily [12.23]. Thus the high symmetry for these tetramer complexes results in a loss of the normal anisotropy associated with pairwise interactions.

12.5 Some unusual valence states

Here we examine e.s.r. spectra for certain out of the ordinary species containing transition metal ions. In Section 12.5.1 we see what happens if an electron is removed from a d^0 complex, and in Section 12.5.2 we examine the results of adding an extra electron to a d^{10} complex. Finally, in Section 12.5.3 a few complexes in which the excess electron is apparently confined to one or more ligands are discussed.

12.5.1 Loss of electrons from d^0 complexes

When impurity anions are incorporated into crystals containing similar anions with one less charge, ionising radiation readily removes an electron and the resulting centre remains quite stable in the crystal. For example, $VO_4^{3-} \rightarrow VO_4^{2-}$ and $PO_4^{3-} \rightarrow PO_4^{2-}$ when in $CaMoO_4$ or $CaWO_4$ crystals. The former, vanadate ion, has a d^0 configuration and the electron is lost entirely from oxygen non-bonding orbitals. (The 'hole' would be delocalised onto all four ligands if the symmetry were strictly tetrahedral, but is largely confined to one, two or three ligands for the normally distorted anions.) Thus the structures that result closely resemble those for ions like PO_4^{2-} (Section 7.6). The major structural difference is that the σ-bonding for PO_4^{3-} is essentially sp^3 on phosphorus but largely a mixture of s and d on vanadium. Also π-bonding is very important for VO_4^{3-} because of the 'empty' d manifold, but probably trivial for PO_4^{3-}.

The e.s.r. results [12.24] for the two radicals PO_4^{2-} and VO_4^{2-}, and indeed for a variety of other comparable species, are remarkably similar. Small positive values for Δg accord with the spin being confined to oxygen, whilst the nearly isotropic hyperfine coupling constants (^{31}P, ^{51}V) are best explained in terms of spin-polarisation of the σ-bonding electrons. In that case, the results suggest that this effect is comparable for PO_4^{2-} and VO_4^{2-}, and this, in turn, suggests that the % s-character is comparable. If the σ bonding in PO_4^{2-} is still *ca.* sp^3 at phosphorus then that at vanadium must be *ca.* d^3s to explain these results. In view of the controversy about σ bonding in ions such as VO_4^{3-}, this is an important result.

12.5.2 Gain of electrons by d^{10} complexes

In most cases, an excess electron is added into a combination of the metal outer s-orbital and ligand σ-orbitals. Typical examples include the formation of Ag^0 from Ag^+ in KCl or in aqueous glasses and of Cd^+ and Hg^+ from Cd^{2+} and Hg^{2+} discussed in Section 7.2. However, the large, isotropic, hyperfine coupling characteristic of such species is not always detected, an interesting example being that of electron addition to linear $Ag(CN)_2^-$ complex anions [12.25]. This gave a species having $A_{\parallel}(^{109}Ag) = 60\,G$ and $A_{\perp}(^{109}Ag) = 30\,G$, with g-values very close to that of the free-spin. It is certainly an electron excess centre, $Ag(CN)_2^{2-}$, and the electron is clearly not largely in the silver $5s$ orbital. We have suggested an orbital comprising $5s$ and $4d_{z^2}$ on silver, together with a considerable contribution from the anti-bonding cyanide σ-orbitals, but this is not definitive at present, and there may well be other possible explanations.

Another interesting example, in which silver(I) does not retain an excess electron, if found when silver atoms react with ethylene. In this case the coupling to ^{109}Ag was even smaller, and the authors suggested that it was accommodated in a hybrid of $4d_{xy} + 5p_x$ on silver [12.26].

Before leaving the topic of electron gain by d^{10} ions, I would like to mention briefly a remarkable result for electron addition to Mn^{2+} ions. These ions, doped into alkali-halide crystals, are high-spin d^5, and one would have expected that the resulting Mn^+ ions would be high- or possibly low-spin d^6.

149

However, these ^6S ions, with the d-manifold exactly half filled are relatively stable, and e.s.r. spectroscopy shows, quite unambiguously, that the excess electron is accepted into the outer, $4s$ orbital on manganese, giving a ^7S ion. The ^{55}Mn hyperfine coupling remains isotropic, but is now dominated by the large, positive contribution from the $4s$ electron, which explains the large value of *ca.* 230 G observed [12.27]. This is perhaps the only example in which an octahedrally coordinated ion with an incomplete d-shell used the outer s orbital for one of its electrons.

12.5.3 *Complexes with paramagnetic ligands*

Mention has already been made of superoxide complexes for which the unpaired electron seems to be largely confined to the O_2 ligand (Section 12.3.4), and $VO_4{}^{2-}$ is another example (Section 12.5.1). There are a number of other such species, and we consider here the anion formed by electron addition to the nitroprusside ion, $(CN)_5FeNO^{3-}$ and various nitroxide derivatives, $L_nM—N(O)R$ (where R is alkyl or aryl).

When nitroprusside salts are irradiated at low temperatures, electron addition seems to occur initially on the N—O group, to give a complex whose e.s.r. parameters indicate that the excess electron is largely confined to one of the π^* NO orbitals (Fig. 12.21(a)).This species can be compared with the superoxide derivatives discussed in Section 12.3.4. The reaction is of considerable interest in that, on annealing above 77 K the electron is transferred irreversibly into the 'd_{z^2}' orbital which is σ^* to the NO ligand. This shift is presumed to be accompanied by a partial or complete loss of the axial cyanide ligand [12.28]. Thus a specific relaxation is needed before the σ^* orbital is deep enough to accept the electron. This is one of the few examples that directly establish Taube's important concept [12.29] that electron-transfer processes are often mediated by conducting ligands.

The technique of 'spin-trapping' which has been mentioned from time to time, and which is extremely useful for the study of reactive species that cannot

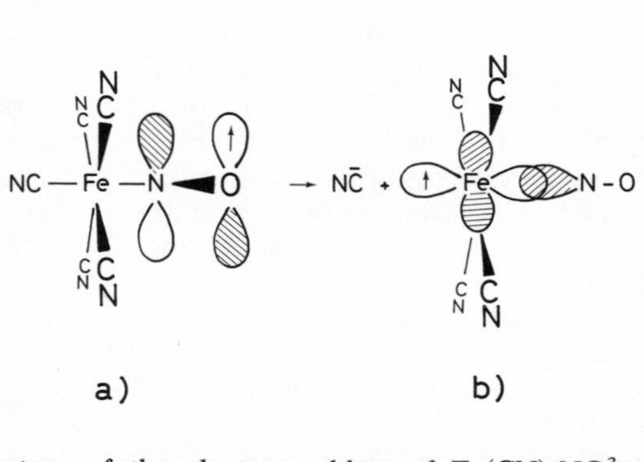

a)   b)

Figure 12.21 Structure of the electron adduct of $Fe(CN)_5NO^{3-}$ first formed (a) at 77 K, in which the electron is essentially confined to the NO ligand, and (b) after warming, in which loss of the axial cyanide allows the electron to move into the d_z^2 (σ^*) orbital.

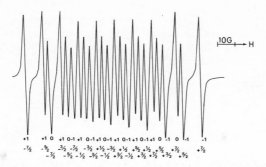

Figure 12.22 First derivative X-band e.s.r. spectrum assigned to $Co(CN)_5$—$N\dot{O}(Ph)$ radicals.

be detected directly by e.s.r. spectroscopy, can be used in the transition-metal complex field also:

$$L_5M\cdot + RNO \longrightarrow \begin{matrix} R \\ \diagdown \\ NO\cdot \\ \diagup \\ L_5M \end{matrix} \qquad (12.4)$$

To be successful, one probably needs a complex with a deficiency of ligands (as in L_5M), and also one that does not have a high electron-affinity, so that the π^* (NO) electron does not move into the d-manifold.

Probably the first metal nitroxide of this type was studied by Swanwick and Waters [12.30] who reacted pentacyanocobaltate(II) ions $[(Co(CN)_5{}^{2-}]$ with various aromatic nitroso compounds. A typical liquid-phase spectrum is given in Fig. 12.22. The data, summarised in this figure, accord well with the simple formulation given above.. However, the small isotropic hyperfine coupling to ^{59}Co (*ca.* 10 G) could arise because of a large positive contribution from the $4s$ orbital occupancy, so it seemed important to check the solid-state spectrum as well [12.31]. This confirmed that the spin-density on cobalt is extremely small, and indeed, the hyperfine coupling can be explained entirely in terms of a σ-spin-polarisation mechanism.

A similar species was prepared by generating $\cdot Mn(CO)_5$ in the presence of RNO molecules, the e.s.r. results showing clearly that addition to give a nitroxide radical had occurred [12.32]. This seems to be a powerful method for detecting unstable transition metal fragments which often give no well defined e.s.r. spectra on their own.

151

CHAPTER 13

Some biological systems

I have endeavoured to introduce examples drawn from biology throughout this book, but these leave some areas largely untouched, and also the reader who is strongly oriented towards biology may find it difficult to locate these examples. In this chapter I therefore list these examples for easy location, and briefly fill in some of the gaps.

13.1 Index of biological topics

There are four major applications of e.s.r. spectroscopy, namely: (A) intrinsic signals from radicals, (B) intrinsic signals from transition metal complexes, (C) signals from irradiated materials, and (D) signals from nitroxide 'spin-labels'. Examples of B, C and D are as follows:

(B) Section 12.3.2 on low-spin ferric haem complexes
Section 12.3.4 on vitamin B_{12}
Section 12.3.6 on Cu(II) proteins
Section 12.4.4 on met-haemoglobin
Section 12.4.6 on metal-ion clusters

(C) Section 12.3.4 on irradiated oxy-haemoglobin
Section 10.4 on the photosynthetic process

(D) Section 9.3 on the solvation of nitroxides
Section 11.4.1 on di-nitroxides

13.2 Intrinsic radical signals from biological materials

It is probable that relatively stable, large organic π-radicals are important intermediates in many biological reactions, especially those involving electron transfer. However, in general, these species are present only in very low concentrations, and an examination of the literature (which is extensive) leaves one with the feeling that the weak, broad singlets that are usually detected in the free-spin area are relatively uninformative. Although one can guess from

152

the systems what types of radical may be involved, e.s.r. is of little use in confirming such guesses.

One type of semiquinone that undoubtedly plays an important rôle in one-electron redox reactions is the flavin semiquinones [13.1]. Flavoproteins, which play an important rôle in many redox reactions catalysed by enzymes, contain sub-units based on the isoalloxazine unit, which readily add one or two extra electrons. The e.s.r. signal from fluid solutions of the free semiquinones

isoalloxanine

are rich in hyperfine structure and identification is unambiguous, but unfortunately one electron reduction of metal-free flavoproteins themselves gives only broad singlets at $g = 2.0032$. There can be little doubt about the identification, but no structural detail can be deduced. Clearly, the semiquinones must all be tightly bound to the slowly rotating polymer. For detailed reviews of intrinsic e.s.r. signals, see ref. 13.2.

13.3 Signals from irradiated materials

Ultraviolet photolyses have been largely confined to photosynthetic systems (see Chapter 10 and ref. 13.3). These studies are clearly of direct biological significance. However, X-ray and γ-ray studies, which have been largely concerned with isolated small molecules with biological connections, have much less direct relevance. They are numerous, and generally conform well to chemical expectation. Examples include aminoacids, sugars, phosphate esters, uracil derivatives, etc. These materials have been studied as single crystals, powders, and dilute solutions in fluid and solid states. Marked differences in behaviour are often noted on going from the pure solids to solutions, and this alone justifies the use of rigid glasses despite the fact that the resulting spectra may be relatively poorly defined.

One long sought after radical which was first detected in certain irradiated biological samples is the alkoxy radical, RO·. Box and his co-workers [13.4] found a strongly anisotropic signal with one very high g-value and extremely large proton hyperfine coupling constants in certain irradiated ribose derivatives. There can be no doubt of the identification as an $RCH_2O·$ radical, since the form of the g-tensor, the large coupling to two inequivalent β-type protons and the directions of the g-tensor components relative to the hydrogen-bond framework in the crystal are all as expected for such radicals. What is very surprising is that despite many attempts to study such radicals in simple monohydric alcohols, no authentic detection had previously been reported (though many obviously incorrect e.s.r. assignments are to be found in the literature). As is indicated in Fig. 13.1, it is necessary to have very precise hydrogen bonding if well defined features are to appear because of the extreme

153

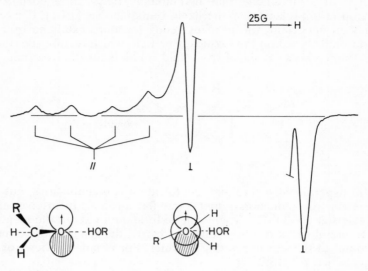

Figure 13.1 First derivative X-band e.s.r. spectrum obtained from thymidine after exposure to ^{60}Co γ-rays at 77 K, showing features assigned to $RCH_2O\cdot$ radicals. The way in which precise hydrogen bonding controls the form of the spectrum is also shown.

sensitivity of the β-proton coupling to the direction of such bonding, and also of the g_z value on the strength of the hydrogen-bonding. Any smearing of these parameters and the spectrum would rapidly broaden.

Less work has been done on whole bio-systems, and obviously the results are usually too complex to interpret. A good example in which well defined signals were obtained is that of irradiated blood cells (Section 12.3 and ref. 13.5). Another well defined signal that is often obtained is that for the 'thymine radical'. This is an 8-line spectrum (Fig. 13.2) assigned to the radical:

There are two reasons why this radical is so frequently detected. One is because of the large coupling to the CH_2 protons (*ca.* 38 G each) (which arises because of the conformation indicated which gives good σ–π overlap) and to the methyl protons, both of which, being β-protons, give almost isotropic interactions, and the other is that electron transfer to the thymine base seems to be rapid even at 77 K. (This assumes that the radical is obtained from the

154

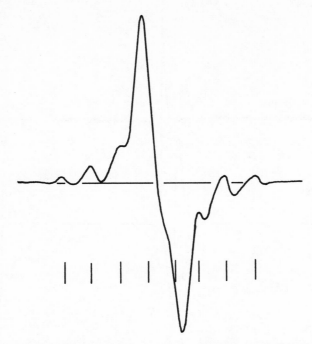

Figure 13.2 First derivative X-band e.s.r. spectrum for a γ-irradiated DNA sample, which shows the 8-line spectrum for the 'thymine radical' together with a broad central doublet.

anion, by protonation.) This result is of considerable importance. It seems that the base molecules, in DNA for example, are so stacked that excess electrons can jump from one to another extremely rapidly, and that the electron-affinity of thymine bases is greater than those of the majority or even all of the others. (I stress, however, that the importance of thymine may be exaggerated just because of its well defined e.s.r. spectrum.)

13.4 Spin-labels

This technique is not concerned with radicals *per se*, but with solvation, conformation and correlation times (see Section 9.3). The e.s.r. technique is important because it can probe times in the 10^{-6}–10^{-11} s region, which is of great importance in many biological systems, and is not easily probed by other techniques. Biopolymers in aqueous solution behave, in many ways, as solids since they move slowly on the 'e.s.r. time-scale'. Hence the e.s.r. spectra of nitroxides chemically attached to such polymers might be expected to resemble those of solid nitroxides, and this is indeed the case in many instances. However, if they are attached to groups that can rotate or librate about some axis relative to the main molecule, their spectra may be partly averaged or may exhibit a 'slow tumbling' form that can be analysed to give a mean correlation time (see Figs. 13.3 and 13.4 and ref. 13.6). (e.s.r. spectra for a typical

155

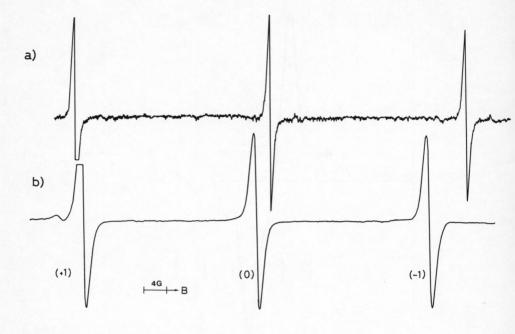

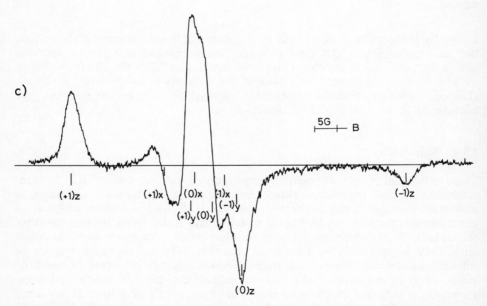

Figure 13.3 First derivative X-band e.s.r. spectra for $(Me_3C)_2NO\cdot$ radicals (a) in water at 20°C; (b) in dodecane at 20°C and (c) in CD_3OD at 77 K. (In (c) $(CD_3)_3CNO\cdot$ radicals were used to give better definition.)

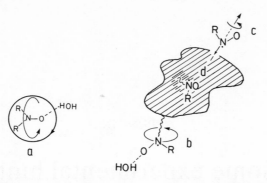

Figure 13.4 (a) Rotation (in water) usually too fast to affect line-widths significantly. $A(^{14}N)$ = normal value in water (≈ 17 G). (b) Attachment to biopolymer by a chain will leave the nitroxide group relatively free to rotate, but slight asymmetry in rate about x, y and z may be introduced. $A(^{14}N) \approx 17$ G. (c) Direct 'surface' attachment with seriously restricted rotation, though it may still be fairly rapid about the C—N bond direction (α). This will lead to a partial averaging of the parameters and a modified solid-state spectrum will result. $A(^{14}N)$ is probably still *ca.* 17 G. (d) The NȮ group is now buried within the polymer, so a 'non-rotating' solid-state spectrum will result. Since $R_2NO\cdot$ molecules are only weakly basic there will be little tendency to drag water into the cavity and so $A(^{14}N)$ will be < 17 G, and probably *ca.* 15–15.5 G unless some other acidic protons are conveniently placed within the cavity.

nitroxide in fluid and solid solutions are given in Fig. 13.3.) In addition to giving some idea of the local movements of nitroxides incorporated into biological polymers, the value for $A_{iso}(^{14}N)$ can, under favourable circumstances, give useful information about the environment of the NO group, specifically about the extent of hydration. As stressed in Section 9.3, since $R_2\dot{N}O$ molecules are only weakly basic, they do not hold water molecules strongly, and hence will not drag water with them into normally non-aqueous regions of biological systems.

Nitroxides have been used with considerable success in the study of membranes [13.7], micelles [13.8] and proteins [13.9]. The reader is referred to the literature for details. Most arguments are based upon systems such as that indicated in Fig. 13.4. Recently, dinitroxides (Section 9.4.1) have also been introduced and shown to be particularly useful in the study of lipid bilayer membranes. See, for example, ref. 13.10.

Some experimental hints

This appendix comprises a motley collection of facts and suggestions that may help student e.s.r. spectroscopists—I'm sure I have omitted various vital pieces of information, but I have done my best to cover most of the snags. I consider (1) the sample, (2) the spectrometer and (3) the spectrum.

A1.1 The sample

(a) If you are studying a single crystal, study the powder also. This often gives a useful check on the validity of the crystal data, and it is, in many cases, easy to make mistakes!

(b) Conversely, if you are studying a powder sample, and crystalline material can be obtained, take some 'snap-shot' views of the crystal—it may reveal much more detail than was available from the powder, and it may show that the powder is unable to give the desired parameters.

(c) Is your powder sufficiently fine? The presence of small crystallites may give rise to multiple small features that can be very misleading. Try rotating the sample if in doubt.

(d) Are your samples rare and therefore only available in small quantities? Use a Q-band spectrometer to enhance sensitivity.

(e) Are the features intense but broad? Try further dilution to avoid spin-spin broadening.

(f) If you are studying liquid-phase samples, have you removed dissolved oxygen? This, like too high a concentration, can lead to broadening because oxygen is paramagnetic.

A1.2 The spectrometer

(a) Are your lines broad? Perhaps the power is too high and you are partially saturating the sample. Always use the lowest possible power. Also, you may be over modulating—this gives loss of resolution so cut down the modulation until no further resolution is apparent.

(b) Is your spectrum too noisy? Use a longer time constant, or a CAT (computer of average transients) which adds coherent signals only.

158

A1.3 The spectrum

(a) Are certain predicted features absent or unduly broad? Try heating and cooling the sample in case some exchange process is giving rise to selective line-broadening.

(b) Have all the features been detected? The analysis of complex spectra (especially in the liquid phase) is very difficult if the first and last features are missing—and it is very easy to miss these if there are a number of equivalent nuclei present because of the binomial distribution of intensities. I strongly recommend that you begin your analysis with the first line, taking distances from this to successive features as hyperfine splittings or multiples thereof. But you MUST have the first line, so 'blow-up' the wings to make sure.

(c) Have you missed features from radicals containing isotopes in low abundance, or even features from another radical? A common mistake is to concentrate on intense features (usually in the $g = 2$ region) and forget to look in the low- and high-field regions. These should always be scanned at high gain, just in case. (For example, in all the early work on radicals in irradiated mercury, tin and lead compounds, the very high field satellite lines from radicals containing ^{199}Hg, ^{201}Hg, ^{119}Sn and ^{207}Pb isotopes were missed, and hence spectral identification was largely guess-work.)

(d) Do you find it impossible to be sure about the correct analysis of a powder spectrum? Try comparing X- and Q-band spectra. The increased separation between features with different g-values at Q-band must fit in with your analysis, and generally only a single scheme will accommodate both spectra. For more details on powder spectra, see P. C. Taylor, J. F. Baugher and H. M. Kriz, *Chem. Rev.*, 1975, **75**, 203.

(e) Are you converting hyperfine data in G to data in MHz or cm^{-1}? Don't forget to multiply by g_e/g_{exp}.

(f) If you have a powder spectrum, can you be sure that g and A are co-linear? If they are not, you really need a single crystal. Remember that the powder spectrum will show turning-points which will be dominated by whichever has the largest field variation, i.e. if ΔA is large and Δg small, you will get close to the correct A values but the g-values will correspond to those for these A values rather than to the principal g-values. (Sometimes this can lead to an extra turning point and hence an extra feature in the powder spectrum that has no specific significance.) If you change to a Q-band spectrometer the form of the spectrum may well change, because Δg is now *ca.* 3.5 times larger, and could begin to compete with ΔA.

(g) Has a solid-state spectrum given unexpectedly small hyperfine anisotropies for your radical? Then try cooling. Frequently radicals, especially if they are small, will librate about certain axes, giving partial averaging, or may undergo complete rotations about certain axes, which will give rise to misleading parallel and perpendicular features. (Some trapped radicals, such as $\cdot CH_3$ or $\cdot NH_2$ have such high rotational energies that they are difficult to obtain in a completely *rigid* form.)

(h) Do certain narrow features appear as triplets with a splitting in the region of 5 G at X-band? Try reducing the power. The first and last features may be 'forbidden' lines resulting because of weak, dipolar coupling to neighbouring magnetic nuclei (usually protons). This coupling can lead to concurrent nuclear transitions (*cf.* Appendix 2) which appear as satellites to the main line. You can diagnose this because the splitting increases with field, and the intensities increase with power.

(i) Do you have unexpected shoulders and peaks to high and low field of $g = 2$? They may result from pair-trapping of radicals (Section 11.1). They are usually accompanied by a narrower feature at $g \approx 4$, and frequently take the form of a quintet, if you include a central line at $g = 2$ as part of the spectrum. Such pairs, formed in photolysed or γ-irradiated solids at low temperature, are usually lost preferentially on annealing.

(j) Is the spectrum too complicated to interpret—or are the features broad and gaussian in shape? Then, if you have it, use an ENDOR spectrometer (Appendix 4). This greatly reduces the number of features, and can pick out hyperfine features in e.s.r. lines that are quite unresolved.

(k) If you have two radicals, one with narrow features and one with broad features, try varying the microwave power. Possibly one saturates more readily than the other, so that at really low power, the narrow, easily saturatable spectrum may well dominate, whilst at high powers the broad spectrum that is difficult to saturate should dominate.

Extracting data

A2.1 Corrections

The full energy equation (Hamiltonian) that describes the resonance conditions must contain *all* possible interactions even though the highly simplified equations that emerge in the high-field approximation, utilised in Chapters 2–4, are often satisfactory. The full equation should contain several terms besides the normal terms for the interaction between the electron and the field (the electron 'Zeeman' term) and the interaction between the electron and the nuclei (the hyperfine term). These include terms for interaction between the field and the nuclei (the nuclear Zeeman term) the nuclei on each other (the spin-spin term of n.m.r. spectroscopy) and a nuclear quadrupole term.

When the hyperfine term is large compared with the applied field, the zero-field coupling between the electron and the nuclei (Chapter 3) must be taken into consideration by using the Breit–Rabi and related equations. This is considered briefly in Section A2.2. If the hyperfine coupling is very large, some transitions become unattainable at normal X- or Q-band frequencies. However, nuclear transitions analogous to normal n.m.r. transitions, can then sometimes be detected, and one must be careful not to confuse these with normal e.s.r. transitions, or incorrect parameters can be derived. These are also discussed in Section A2.2.

In contrast, when the nuclear moments are relatively large but the hyperfine coupling relatively small, the applied field tends to orient the nuclei directly. This encourages transitions in which nuclei also change orientation with the electron. Such transitions are discussed in Section A2.3. The presence of a large electric field gradient tends to keep quadrupolar nuclei aligned to the molecular frame. This also facilitates 'forbidden' transitions and can totally alter the appearance of e.s.r. spectra. Such transitions are significant when the hyperfine term is relatively small, and the electric field-gradient and quadrupole moment are large (Section A2.4).

When g-values deviate far from 2.0023 as a result of orbital magnetism, this magnetism also gives a contribution to the hyperfine coupling constants. Thus, if you wish to interpret the hyperfine components in terms of orbital populations, they must first be adjusted to the values that would have been obtained at $g = 2.0023$. This is discussed in Section A2.5.

It is not my concern to delve into these extra effects in depth. All I attempt to do is to make the reader aware of the pitfalls, and to give suitable references to the detailed theory, which seems to be well understood.

A2.2 Corrections when A is large $(S = \frac{1}{2})$

A2.2.1 *Isotropic*

(*Cf*. Fig. 3.2 for $I = \frac{1}{2}$ and Fig. A2.1 for $I = \frac{3}{2}$.) Note: At low fields the quantum number F is required, where $F = I \pm \frac{1}{2}$. Thus, for $I = \frac{1}{2}$, $F = 1$ or 0. Then m_F, the z component of the total angular momentum, ranges between $\pm F$. Thus for $F = 1$, $m_F = \pm 1, 0$. At zero field, there are always two levels $(I \pm \frac{1}{2})$, separated by the zero-field splitting, $(I + \frac{1}{2}) g \beta A$ (MHz).

In Fig. 3.2 the way in which the transitions are shifted to low field is indicated. In Fig. A2.1, for $I = \frac{3}{2}$, the thick vertical lines indicate the normal e.s.r. transitions. From both figures it is clear that as the relative energy of the microwave field falls, so the lower field transition moves rapidly to zero-field and then becomes unattainable. It is replaced, however, by an 'n.m.r.' transition indicated as NMR 1 in Figs. 3.2 and A2.1. Other possible 'n.m.r.'

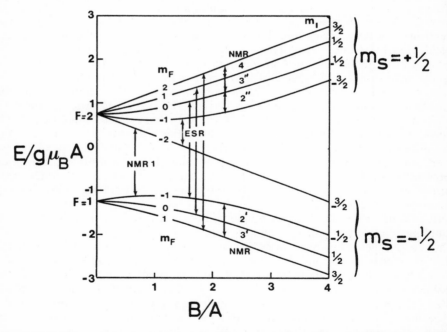

Figure A2.1 Energy levels for a radical having $S = \frac{1}{2}$ and $I = \frac{3}{2}$. In zero field the combined quantum number F must be used, the two values being $\frac{3}{2} + \frac{1}{2} = 2$ and $\frac{3}{2} - \frac{1}{2} = 1$. The $F = 2$ level splits into $+2$, $+1$, 0, -1 and -2 components in a magnetic field, and the $F = 1$ level splits into $+1$, 0 and -1 components. These become the normal M_I components in high field. The e.s.r. transitions are shown for a fixed value of B. Extra 'n.m.r.' transitions are also indicated. (Taken from ref. A2.1 with permission.)

transitions are also indicated. These transitions only involve a nuclear reorientation, hence the n.m.r. label. The way in which these resonances move as a function of the hyperfine coupling is indicated in Fig. A2.2. Note the relative insensitivity of the $M_I = -\frac{1}{2}$ e.s.r. transition, which never moves outside the range of an X-band e.s.r. spectrometer. Clearly, however, if A is very large, this transition alone, especially if it is broad, may not give a very accurate value for the hyperfine coupling. However, the NMR 2 transition is then nearby, and this can be used to check A, or to give the g-value as well of this unknown. The $+\frac{1}{2}$ e.s.r. line is lost when $A = B_0$. The NMR 1 line then takes over, but only gives a detectable signal for quite a small range of A-values. The two NMR transitions occur at the field values

$$B = \pm B_0\{[(2I+1)A - 2B_0]/(2B_0 - A)\} \tag{A2.1}$$

the positive sign giving NMR 1 and the negative, NMR 2.

To calculate A_{iso} and g_{av} from the field values of the observed e.s.r. transitions, the Breit–Rabi equation (A2.2) is used:

$$E^{\pm}/(g\mu_B A) = -\frac{1}{4} \pm \frac{(2I+1)}{4}\left\{1 + \frac{4xm_F}{(2I+1)} + x^2\right\}^{1/2} \tag{A2.2}$$

(A is in G, the nuclear Zeeman term is neglected, $x = 2B/A$ $(2I+1)$ and m_F is the z-component of the total angular momentum F $(= I \pm \frac{1}{2})$.) This gives, for the $M_I = \pm I$ transitions for any value of I,

$$A = \frac{2B_0(B_2 - B_0)}{B_0(2I+1) - B_2} = \frac{2B_0(B_0 - B_1)}{B_0(2I+1) - B_1} \tag{A2.3}$$

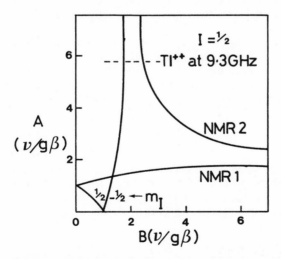

Figure A2.2 The lines give field values for various transitions for a radical with $S = \frac{1}{2}$ and $I = \frac{1}{2}$, as a function of the isotropic hyperfine coupling, A, (in units of $\nu/g\mu_B$). As an example, the position of the Tl^{2+} ion is indicated. (Taken from ref. A2.1 with permission.)

when B_0 is the resonance corresponding to the g-value, B_2 is the high-field transition and B_1 the low-field transition. Thus, for a radical with $I = \frac{1}{2}$,

$$A = \frac{2B_0(B_2 - B_0)}{2B_0 - B_2} = \frac{2B_0(B_0 - B_1)}{2B_0 - B_1} \tag{A2.4}$$

Expressions for the inner lines when $I > \frac{1}{2}$ are more complicated. However, when the shifts are relatively small, which is the situation most commonly encountered, 'corrected' line positions can be very easily derived using the triangle of numbers shown in Scheme A2.1. (This triangle is readily con-

SCHEME A2.1

Number triangle giving the shifts to second-order for each component of a hyperfine multiplet from a single nuclear spin I. The shifts are in units of $A^2/(4g\mu_B B)$ (MHz).

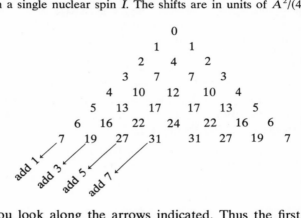

structed if you look along the arrows indicated. Thus the first set is derived from the zero by adding 1, the second from 1 by adding 3, the third from 2 by adding 5, etc.)

If, as sometimes happens, only the $M_I = -I$ (high-field) line can be detected, one can guess that $g \sim 2$, and hence calculate A using Equation A2.3 or if there are radicals containing non-magnetic isotopes these will give the correct g-value. Again, if one of the n.m.r. transitions can be detected this gives enough information for both A and g to be derived.

A2.2.2 *Anisotropic*

The simple Breit–Rabi equation can still be used for anisotropic spectra, provided the anisotropy is small ($2B \le 100$ G for $A > 2000$ G). Unfortunately, it is no longer applicable if the hyperfine anisotropy is large, and the equations that are obtained are such that an iterative technique is needed. In the case of axial symmetry and $I = \frac{1}{2}$,

$$A_\parallel = \frac{2H_0(\parallel)[H_0(\parallel) - H_2(\parallel)] + \frac{1}{2}(A_\parallel^2 - A_\perp^2)}{H_2(\parallel) - 2H_0(\parallel)} \tag{A2.5}$$

$$A_\perp = \frac{2H_0(\perp)[H_0(\perp) - H_2(\perp)] + \frac{1}{2}A_\perp^2 - \frac{1}{8}(A_\parallel + A_\perp)^2}{H_2(\perp) - 2H_0(\perp)} \tag{A2.6}$$

164

A superior and quite general method has been given by Byfleet *et al.*, which is recommended if $I > \frac{1}{2}$ and the anisotropy is relatively large (A2.3).

In general, A values need to be corrected by this method for values $> ca.$ 400 G, but g values should be corrected if $A > ca.$ 100 G, unless very precise values are required.

A2.3 Equivalent nuclei when A is large

A2.3.1 *Isotropic*

Throughout this text we have assumed that the methyl radical gives rise to a 1:3:3:1 quartet in fluid solution. In fact, under very high resolution conditions, if the lines were sufficiently narrow, the two inner features would split into $(1+2)$, $(1+2)$ components. When A is as small as 23 G this is rarely observed, since it is necessary that $A^2/2B_0 > \Delta B$ (the linewidth of the e.s.r. transition). However, it can cause an asymmetry and apparent loss in intensity for the two inner lines. It becomes rapidly more important as A increases. Why should this occur? The key lies in the electron + nuclear coupling and zero-field splitting. The situation is not quite as simple as that for nuclei with the same net spin: for example ^{14}N ($I = 1$) is not exactly the same as, say, $H_2\dot{C}O^-$ ($I = 1$). This is because, in the latter case, $I = 0$ is an alternative arrangement not available to ^{14}N. The situation for two equivalent ($I = \frac{1}{2}$) nuclei is summarised in Fig. A2.3. This is essentially a combination of the diagram for the behaviour of radicals with $I = 0$ (see Fig. 1.1) and $I = 1$. At high-field, lines $(2+3)$ and $(6+7)$ combine to give the two familiar '$M_I = 0$' lines, but at low field these split out to give doublets. The relative shifts are still given by the triangle in Scheme A2.1, being 0 for the $I = 0$ component and 2, 4, 2 for the $I = 1$ triplet (in units of $A^2/4B$).

For 3 equivalent ($I = \frac{1}{2}$) nuclei, we get combined ($I = \frac{3}{2}$) and ($I = \frac{1}{2}$). At zero-field these behave separately, giving $F = 2$ and 1 for one set and $F = 1$ and 0 for the other set. These diverge in the usual way in a magnetic field, and eventually the four lines from the $I = \frac{1}{2}$ radicals merge with the inner four lines from the $I = \frac{3}{2}$ radicals and the spectrum becomes the familiar 1:3:3:1 quartet. At low-fields it becomes a 1:(1+2):(1+2):1 sextet as indicated in Fig. A2.3. You can think of the 1:1:1:1 quartet as belonging to the $I = \frac{3}{2}$ set, and the 2:2 doublet as belonging to the $I = \frac{1}{2}$ set, and the shifts for these sets are again given in Scheme A2.1. Note that the doublet is 2:2 because there are two ways of combining the three nuclear spins after subtracting the $I = \frac{3}{2}$ quartet. In exactly the same way for four equivalent nuclei (as in SF_5 for example) we get radicals with $I = 2$ (relative intensity 1), with $I = 1$ (relative intensity 3) and with $I = 0$ (relative intensity 2). Hence the spectrum indicated in Fig. A2.3 is obtained, the shifts corresponding to these I-values being given in Scheme A2.1. These schemes work very well provided A is not too large.

A2.3.2 *Anisotropic hyperfine coupling*

This relatively simple pattern no longer holds if the hyperfine coupling is anisotropic—again as might be expected, since the zero-field coupling pattern

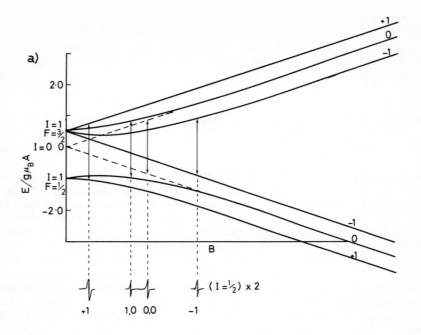

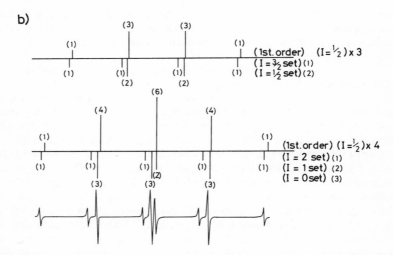

Figure A2.3 (a) Energy levels for a radical containing two equivalent nuclei with $I = \frac{1}{2}$, showing how two sets of levels are generated, one from radicals with net $I = 1$, giving the $F = \frac{3}{2}$ and $\frac{1}{2}$ zero-field levels, and the other with net $I = 0$ (dashed lines). e.s.r. transitions are shown as vertical lines, and the resulting spectrum is indicated. (b) Stick diagrams for the situation that is found for 3 equivalent nuclei with $I = \frac{1}{2}$, giving two sets of transitions ($I = \frac{3}{2}$ and $I = \frac{1}{2}$) and that for 4 equivalent nuclei showing three sets of transitions. (A first derivative spectrum is also indicated in this case, such as that observed for $\cdot SF_5$ radicals.)

166

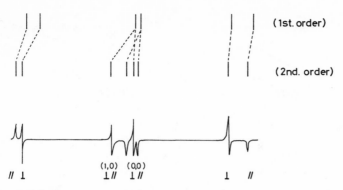

Figure A2.4 Stick diagram and first derivative spectrum for a hypothetical radical containing two equivalent nuclei with $I = \frac{1}{2}$ and an axial hyperfine coupling tensor. Note the parallel shifts and central line splittings are smaller than those for the perpendicular components.

also changes, as discussed above. Considering only the case for a pair of equivalent nuclei (as, for example, in $H_3P\!-\!PH_3^+$ discussed in Section 7.5) we have spectra of the type shown in Fig. A2.4. To second order the splitting for the '$M_I = 0$' lines into the 1, 0 and 0, 0 components is given by

$$\Delta_{\parallel} = A_{\perp}^2/B \quad \text{and} \quad \Delta_{\perp} = (A_{\parallel}^2 + A_{\perp}^2)/2B \tag{A2.7}$$

Thus the parallel splitting is less than the perpendicular, and the difference increases rapidly as $A_{\parallel} - A_{\perp}$ increases. Hence for radicals with a large $A_{\parallel}$ and small $A_{\perp}$, $\Delta_{\parallel}$ can be ignored, but $\Delta_{\perp}$ may be appreciable (e.g. F_2^-).

A2.4 Corrections when A is small

In most radicals, the electron-nuclear hyperfine energy is greater than the nuclear Zeeman term, which is neglected. However, this Zeeman term, $g_N\mu_N B$ is 5.5 G at X-band for protons, and hence for typical α- and β-proton hyperfine couplings it has a significant effect. This takes the form of two extra transitions, which involve both the nucleus and the electron ($\Delta M_I = \pm 1$; $\Delta M_s = \pm 1$). These transitions are weak for $A(^1H) \sim 20$ G, but as $A(^1H)$ falls, they become more allowed.

In Fig. A2.5 the types of transitions are indicated for a case in which $A > 2g_N\mu_N B$ ($> 2\nu_N$) (the unequal splittings of the two levels is caused by the nuclear Zeeman term). The two inner transitions involve a nuclear reorientation *as well as* an electron reorientation. We have previously assumed that such simultaneous nuclear spin-inversions are forbidden. This is true when $A \gg \nu_N$, and remains true for all values of A when the applied field is directly along one of the three principal directions of the hyperfine coupling tensor. When the field is off these directions however, and $A \approx 2\nu_N$, these lines become quite intense. The situation is summarised approximately in Fig. A2.6. Note that the extent of anisotropy is an important factor, both in controlling line positions, and also intensities.

167

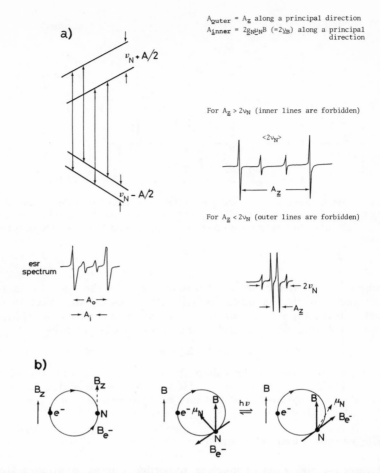

Figure A2.5 (a) The four possible transitions for a radical having a single coupled nucleus with $I = \frac{1}{2}$. In this example, $A > 2g_N\mu_N B$ and hence the two inner transitions involve simultaneous nuclear transitions and are formally 'forbidden'. (Note. These are not 'n.m.r.' transitions of the type discussed in A2.2. Also note that the unequal splitting of the nuclear levels arises because of the nuclear Zeeman interaction: which splitting is greatest depends on the relative signs of the two terms.) (b) Shows how for B along a principal direction, B_{e^-} at N is colinear with B_z, but for B along an arbitrary direction, μ_N aligns along a resultant field, and this direction changes during the electron transition.

What are the experimental consequences? If you are deriving the hyperfine tensor components from a set of rotations about arbitrary axes that bear no relation to the principal directions, due note of all four transitions must be taken. Errors can be appreciable if only the most intense lines are monitored, when $A \gtrsim 2\nu_B$. If only powder spectra are available, all will be well provided the turning points correspond to the principal values of the hyperfine coupling

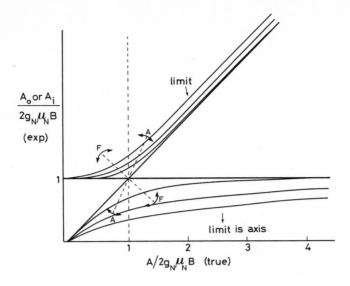

Figure A2.6 Plot of the true hyperfine coupling constant as $(A/2g_N\mu_N B)$ against the experimental values as $(A_0/2g_N\mu_N B)$ or $(A_i/2g_N\mu_N B)$. The two straight lines are for the principal tensor directions, and the sets of lines are for progressive deviations therefrom. (*cf.* ref. A2.) The dashed arrows indicate regions of formally allowed (A) and forbidden (F) transitions.

tensor(s). If not, and A is small, you may be 'looking' along a direction for which all four transitions are relatively intense. In that case neither set will give the true hyperfine coupling. The equations given by Poole and Farach provide a simple aid in estimating the correct values [A2.4].

The situation for $I > \frac{1}{2}$ and for sets of equivalent nuclei is obviously more complicated, but again becomes simple along the principal axes. One particular situation is quite frequently encountered, in which two nuclei are coupled, one giving a large splitting and the other a very small or even an unresolvable interaction. The weak coupling is then likely to be largely dipolar, and hence the situation in Fig. A2.6 holds with $A \ll 2\nu_N$. Hence weak satellite lines separated by $g_N\mu_N B$ from the main transitions are detected. A familiar example, first discovered and explained by Tramell *et al.* [A2.5], is that of hydrogen atoms trapped in an aqueous or acidic environment (Fig. A2.7). These lines are formally forbidden and indeed cannot be detected at low microwave powers. However, at high powers they grow increasingly more intense. Note that the splitting is smaller for the low-field line than for the high-field line because of the dependence of ν_N on the magnitude of the field. The explanation of these lines is that several protons on neighbouring water or acid molecules are close enough to give a significant dipolar coupling, and the satellite lines then correspond to electron transitions occurring together with transitions of one of these protons.

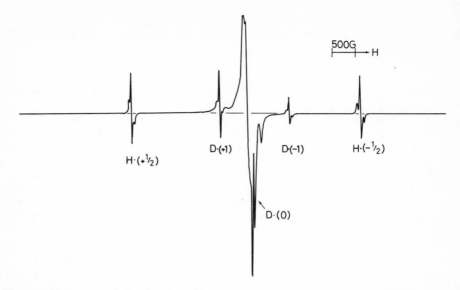

Figure A2.7 First derivative X-band e.s.r. spectrum for hydrogen and deuterium atoms trapped in a γ-irradiated aqueous ($H_2O + D_2O$) sulphuric acid glass. This shows the 'spin-flip' satellite features which come from protons close to the H· or D· atoms. The central features are for ·OSO_3H (or SO_4^-) and ·SO_3^- radicals. The curious intensities for the +1, 0 and −1 components of D· atoms arises because of a cross relaxation between the rapidly relaxing electrons on ·OSO_3H radicals (and probably some ·OH and ·OD radicals) and the slowly relaxing electrons on the atoms. The power needed to show the satellite lines has saturated the atom lines, but cross relaxation has partly desaturated and H· and especially the D· lines.

A2.5 Nuclear quadrupole effects

The way in which the presence of a quadrupolar nucleus in an asymmetric field can affect hyperfine coupling has been outlined in Section 8.1. The examples discussed there were R_2C—Cl and $R_2\dot{C}Br$ radicals, for which very large effects are observed (along the y axis) because the quadrupolar energy then exceeds the hyperfine energy. Such cases have to be treated individually, but are only rarely encountered. The more common situation in which the quadrupole energy is relatively small has been treated in detail, especially by Bleaney (*cf.* ref. A2.6). Some points to bear in mind are:

(i) The quadrupolar interaction takes the form of a traceless tensor.
(ii) The interaction only involves the nucleus, not the electron, and hence the effects are not frequency dependent.
(iii) The noticeable effects are (*a*) small shifts which depend on M_I^3 and hence are symmetrical about the centre of the spectra and (b) the appearance of $\Delta M_I = \pm 1$ or ± 2 transitions. This latter effect (b) is comparable with that induced by hyperfine anisotropy, discussed above, but is usually much larger.
(iv) When H is parallel to the axis of an axial electric field gradient there is no quadrupole effect.

170

Quantitative diagrams showing these changes are given by Abragam and Bleaney [A2.6]. For other useful discussions of quadrupole effects, see refs. A2.7 and A2.8.

A2.6 Corrections for orbital contributions to hyperfine coupling constants

When one or more of the components of the g-tensor are well removed from 2.0023, the induced orbital motion gives rise to an extra component in the hyperfine coupling. Thus the measured coupling is made up of the normal spin-only term together with an orbital magnetic term, and if one wishes to convert the coupling constants into orbital populations in the manner pre-scribed in Section 4, due allowance must be made for this contribution. (Thus, this is not a 'correction' but an adjustment.) It is used to express these adjustments in terms of $\lambda/\Delta E$, when λ is the spin orbital coupling constant and ΔE is the sum of the energies of separation between the half-filled level and those to which it is magnetically coupled. However, Δg, if it is not too large, also depends upon $\lambda/\Delta E$ (Chapters 2 and 12) so, in my view, it is wise to express these 'corrections' in terms of the experimental g-values, thereby avoiding some of the approximations that are otherwise incorporated.

The calculations, to first order in $\lambda/\Delta E$ are quite straight forward and are given in the standard texts. Some useful equations that result for $S = \frac{1}{2}$ systems are as follows:

(i) For a p-orbital system, with $\Delta g \sim 0$ and $|\Delta g_\perp|$ large, use:

$$\left\{ \begin{array}{l} A_\parallel = A_{iso} + 2B \\ A_\perp = A_{iso} - B(1 \pm \tfrac{5}{2}\Delta g_\perp) \end{array} \right\} \tag{A2.8}$$

(ii) In the general p-orbital case for three g-values that differ significantly from 2.00, use:

$$\left. \begin{array}{l} A_x = A_{iso} + 2B(1 \mp \tfrac{5}{4})\Delta g_x \\ A_y = A_{iso} - B(1 \pm \tfrac{5}{2})\Delta g_y \\ A_z = A_{iso} - B(1 \pm \tfrac{5}{2})\Delta g_z \end{array} \right\} \tag{A2.9}$$

(iii) For d-orbitals, a simplified scheme is:

$$\left. \begin{array}{l} A_\parallel = A - 2B(1 \mp \tfrac{7}{4}\Delta g_\parallel) \\ A_\perp = A + B(1 \pm \tfrac{7}{2}\Delta g_\perp) \end{array} \right\} \tag{A2.10}$$

(*Note:* for d_{z^2}, use $+2B$ and $-B$), but the equations become very much more complicated as the Δg values increase.

(*Note:* You must convert from G to, say, MHz before carrying out this calculation for A_{iso} and $2B$. If Δg is negative, use the positive signs, and vice versa.)

A2.7 Links with n.m.r. spectroscopy

It is worth noting that isotropic coupling constants obtained from liquid-phase studies of radicals or complexes having large g-shifts will not be equal to the

'Fermi contact' coupling, and cannot be used directly to calculated s-orbital populations. These must be adjusted to 'free-spin values' in order to obtain the true contact term. This factor is also of importance in n.m.r. studies. As mentioned previously, provided the electron-spin is effectively made to relax rapidly with respect to the nuclei, n.m.r. signals can be obtained from liquid-state samples. The resulting shifts can be used to derive isotropic hyperfine coupling constants, but again, great care should be exercised in extracting structural data therefrom until the contact and 'pseudo-contact' contributions have been derived. These problems are discussed, for example, in refs. A2.6 and A2.9.

APPENDIX 3

ENDOR spectroscopy

A3.1 The experiment

ENDOR stands for electron-nuclear double resonance. Of various forms of double resonance developed as extensions of e.s.r. spectroscopy, this is quite the most important and most commonly used. It has close links with the various forms of double resonance techniques used in n.m.r. spectroscopy. My sole concern here is to indicate what is done and what is thereby revealed. The topic is covered in great depth by Kevan and Kispert [A3.1].

ENDOR spectrometers use a strong radio frequency (n.m.r.) field to induce n.m.r. transitions which are observed as a change in the intensity of an electron resonance transition. Thus the normal microwave field of an e.s.r. spectrometer is increased to give strongly saturating conditions, and this saturation is 'released' by the additional radio frequency field, giving an enhanced signal. The pertinent situation is illustrated for a hydrogen atom in Fig. A3.1. Partial saturation makes the e.s.r. signal weaker than it should be. The magnetic field is held on one of the e.s.r. lines (it doesn't matter which) and the radio frequency is swept between ν_1 and ν_2, which must be equally displaced above or below ν_B. As ν reaches ν_1 or ν_2 extra opportunities for electron spin relaxation are introduced and the e.s.r. signal is enhanced. This enhancement emerges as the ENDOR signal.

A3.2 Advantages of ENDOR

For studying crystals, ENDOR can be used to unravel hyperfine interactions that contribute to the width of an unresolved e.s.r. line. The classic example is Feher's work on F-centres in alkali-halide crystals [A3.2]. In this case, many equivalent nuclei contribute hundreds of features to the e.s.r. spectrum, but since groups of equivalent nuclei give only one pair of lines either side of ν_B in the ENDOR spectrum, the individual contributions were easily resolvable. In liquid-phase studies e.s.r. lines are usually much narrower, and hence all but the weakest interactions may be resolved. However, the resulting e.s.r. spectrum may be extremely complicated, and overlapping features may make analysis, at the best, ambiguous. An example will help to illustrate this.

173

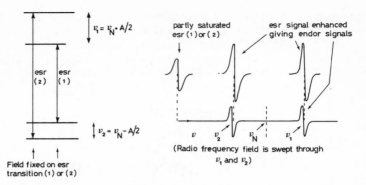

Figure A3.1 The way in which ENDOR transitions come about for a single $I = \frac{1}{2}$ nucleus. ν_B is the normal n.m.r. frequency ($\approx 14\,\text{MHz}$ for a proton at X-band), $\nu_1 = \nu_B + A/2$ and $\nu_2 = \nu_B - A/2$. Since frequency is swept in this experiment, a constant magnetic field is indicated. The key is that as the radio-frequency induces nuclear transitions, the electron is able to relax more efficiently and hence the e.s.r. signal regains strength. This change is recorded as the ENDOR signal.

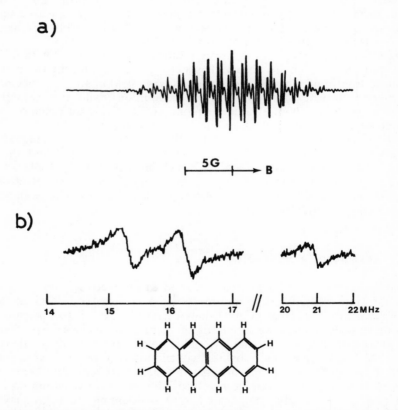

Figure A3.2 The e.s.r. (a) and Ednor (b) spectra for tetracene cations in sulphuric acid.

174

Consider the e.s.r. spectrum of the tetracene cation shown in Fig. A3.2. This comprises, in theory, three sets of five transitions from four equivalent protons thus we expect $5^3 = 125$ lines, some of which overlap fortuitously. However, the ENDOR spectrum comprises only three pairs of lines, well separated from each other, which give the hyperfine coupling constants extremely accurately from:

$$A \text{ (MHz)} = \nu_1 - \nu_2 \tag{A3.1}$$

or

$$A = 2(\nu_1 - \nu_B) \tag{A3.2}$$

APPENDIX 4

Hyperfine coupling constants (G) for unit population of atomic orbitals calculated from Hartree–Fock atomic wave-functions

		$A_{iso}^0(G)$	$2B^0(G)$
Li	6	39.09	
Be	9	−127.46	
B	10	241.41	12.69
	11	720.82	37.90
C	13	1115.4	61.30
N	14	555.1	33.24
	15	−778.7	−46.62
O	17	−1653.2	−102.18
F	19	17 100.3	1 081.06
Ne	21	2 040.8	−131.20
Na	23	223.4[316exp]	
Mg	25	−118.7	
Al	27	980.1	43.19
Si	29	−1 213.2	−61.34
P	31	3 663.7	201.39
S	33	971.2	56.51
Cl	35	1 666.0	100.68
	37	1 386.7	84.07
K	39	51.7[82.5exp]	
	41	28.4	
Ca	43	−148.9	
Ga	69	2 657.7	105.66
	71	3 377.0	134.26
Ge	73	−532.8	−25.15
As	75	3 415.4	178.78
Se	77	4 792.2	269.24
Br	79	7 738.2	457.87
	81	8 340.9	493.53
Kr	83	−1 432.5	87.23
Rb	85	199.4	
	87	675.9	
Sr	87	−165.1	
In	113	3 405.3	148.43
	115	3 412.3	148.73
Sn	115	−6 647.2	−335.99
	117	−7 224.4	−365.17
	119	−7 576.7	−382.98
Sb	121	6 068.3	336.46
	123	3 285.9	182.19
Te	123	−8 051.1	−476.43
	125	−9 706.6	−574.40
I	127	7 294.5	453.18
Xe	129	−11 784.8	−751.04
	131	3 492.3	222.56
Cs	133	349.24	

		$A_{iso}^0(G)$ (outer s)	$2B^0(G)$	$A_{iso}^0(G)$ (inner s)
Sc	45	654.7	−37.78	11 629.5
Ti	47	−177.5	+12.11	−3 259.6
	49	−177.6	+12.11	−3 260.3
V	50			
V	51	947.7	−74.03	18 088.9
Cr	53	−230.1	20.18	−4 572.6
Mn	55	1 127.9	−109.47	23 340.8
Fe	57	163.2	−17.34	3 513.9
Co	59	1 320.2	−152.30	29 539.1
Ni	61	542.7	−67.49	12 603.8
Cu	63	1 764.0	−235.11	42 467.0
	65	1 889.8	−251.88	45 496.3
Zn	67	452.5	64.31	11 278.3
Y	89	−233.8	9.12	−3 317.9
Zr	91	−519.8	24.24	−7 316.4
Nb	93	+1 550.9	−83.01	22 077.4
Mo	95	−459.9	27.65	6 689.0
Mo	97	−469.5	28.23	6 828.7
Tc	99	+1 742.5	−116.04	26 043.0
Ru	99	−376.2	27.48	5 798.5
Ru	101	−421.3	30.78	6 494.4
Rh	103	−284.9	22.66	4 539.2
Pd	105	−398.9	34.34	6 577.7
Ag	107	−417.3[611exp]	38.70	7 127.0
Ag	109	−479.7[706exp]	44.48	8 192.9
Cd	111	−2 317.8	228.47	41 021.5
	113	−2 424.7	239.01	42 912.6
La	175	1 039.9	−43.03	16 161.8
	176	1 110	−47	17 757.4
Hf	177	320	−15.1	4 735.4
	179	−188	+8.9	−2 785.5
Ta	181	1 411.5	−73.90	20 369.7
W	183	534.4	−30.53	7 636.8
Re	185	3 173.8	−195.99	45 348.1
	187	3 206.4	−198.00	45 813.3
Os	189	1 177.8	−78.10	16 937.1
Ir	191	310.1	−22.0	4 510.0
	193	327.5	−23.2	4 762.9
Pt	195	3 692.6	−278.47	54 495.1
Au	197	311.35	−24.91	4 675.8
Hg	199	3 396.7	−285.24	52 010.9
	201	−1 253.6	+105.27	19 194.6

APPENDIX 4 (*Continued*)

		$A_{\text{iso}}^0(G)$	$2B^0(G)$	$A_{\text{iso}}^0(G)$ (outer s)	$2B^0(G)$	$A_{\text{iso}}^0(G)$ (inner s)
Ba	135	469.87				
	137	525.63				
Tl	203	14 838.9	657.36			
	205	14 983.0	663.74			
Pb	207	6 843.9	349.74			
Bi	209	6 371.7	356.59			

[Taken from C. Froese, *J. Chem. Phys.*, 1966, **45,** 1417.]

References

1.1 E. Zavoisky, *J. Phys. USSR.*, 1945, **9,** 211.

1.2 B. Bleaney and K. W. H. Stevens, *Rept. Progr. Phys.*, 1953, **16,** 108.

1.3 J. H. Van Vleck, *The Theory of Electric and Magnetic Susceptibilities*, Oxford University Press, 1932.

1.4 S. I. Wiseman and D. Banfill, *J. Amer. Chem. Soc.*, 1953, **75,** 2534.

1.5 W. Gordy, W. B. Ard and H. Shields, *Proc. Nat. Acad. Sci.*, 1955, **41,** 983.

1.6 J. F. Gibson, D. J. E. Ingram, M. C. R. Symons and M. G. Townsend, *Trans. Faraday Soc.*, 1957, **53,** 914.

1.7 A. Abragam and B. Bleaney, *Electron Paramagnetic Resonance of Transition Ions*, Oxford University Press, London, 1970.

1.8 A. Carrington, *Microwave Spectroscopy of Free Radicals*, Academic Press, London, 1974.

1.9 P. B. Davies, D. K. Russell, B. A. Thrush and F. D. Wayne, *J. Chem. Phys.*, 1974, **60,** 3178.

1.10 P. W. Atkins and M. C. R. Symons, *The Structure of Inorganic Radicals*, Elsevier, Amsterdam, 1967.

1.11 W. A. Waters, *The Chemistry of Free Radicals*, Oxford University Press, 1946.

1.12 L. Kevan and L. D. Kispert, *Electron Spin Double Resonance Spectroscopy*, Wiley-Interscience, New York, 1976.

1.13 C. P. Poole, *Electron Spin Resonance: A Comprehensive Treatise on Experimental Techniques*, Wiley-Interscience, New York, 1967.

1.14 See, for example, E. de Boer and J. L. Sommerdijk, in *Ions and Ion Pairs in Organic Reactions*, Ed. M. Szwarc, Wiley-Interscience, New York, 1972, p. 289.

3.1 T. F. Hunter and M. C. R. Symons, *J. Chem. Soc. (A)*, 1967, 1770.

4.1 H. Zeldes, G. T. Trammell and R. Livingston, *J. Chem. Phys.*, 1960, **32,** 618.

5.1 J. E. Wertz and J. R. Bolton, *Electron Spin Resonance*, McGraw-Hill, New York, 1972.

5.2 N. M. Atherton, *Electron Spin Resonance*, Halsted Press, London, 1973.

6.1 J. A. Brivati, K. D. J. Root, M. C. R. Symons and D. J. A. Tinling, *J. Chem. Soc. (A)*, 1969, 1942.
6.2 H. M. McConnell, *J. Chem. Phys.*, 1956, **24,** 632.
6.3 R. W. Fessenden and R. H. Schuler, *J. Chem. Phys.*, 1963, **39,** 2147.
6.4 R. M. Dessau, *J. Amer. Chem. Soc.*, 1970, **92,** 6356.
6.5 Y. B. Taarit, M. C. R. Symons and A. J. Tench, *J.C.S. Faraday I*, 1977, **73,** 1149.
6.6 I. S. Ginns and M. C. R. Symons, *J.C.S. Dalton*, 1972, 185.
6.7 T. M. McKinney and D. H. Geske, *J. Amer. Chem. Soc.*, 1965, **87,** 3013; G. W. Eastland and M. C. R. Symons, *Chem. Phys. Letters*, 1977, **45,** 422.
6.8 W. C. Danen and R. C. Rickard, *J. Amer. Chem. Soc.*, 1975, **97,** 2303.
6.9 P. J. Krusic, T. A. Rettig and P. von R. Schleyer, *J. Amer. Chem. Soc.*, 1972, **94,** 995.
6.10 M. C. R. Symons, *J.C.S. Perkin II*, 1973, 797.
6.11 A. R. Lyons and M. C. R. Symons, *J.C.S. Faraday II*, 1972, **68,** 502.
6.12 J. E. Bennett and L. H. Gak, *Trans. Faraday Soc.*, 1968, **64,** 1174.
6.13 H. Hefter and H. Fischer, *Ber. Bunsenges, Phys. Chem.*, 1970, **74,** 494.
6.14 B. C. Gilbert and R. O. C. Norman, *J. Chem. Soc. (B)*, 1966, 86.
6.15 T. F. Hunter and M. C. R. Symons, *J. Chem. Soc. (A)*, 1967, 1770.
6.16 N. M. Fox, J. M. Gross and M. C. R. Symons, *J. Chem. Soc. (A)*, 1966, 448.
6.17 M. B. Yim and D. E. Wood, *J. Amer. Chem. Soc.*, 1976, **98,** 2053.
6.18 M. C. R. Symons, R. C. Selby, I. G. Smith and S. W. Bratt, *Chem. Phys. Letters*, 1977, **48,** 100.
6.19 J. P. Colpa and E. de Boer, *Mol. Phys.*, 1964, **7,** 333.
6.20 H. J. Bower, J. A. McRae and M. C. R. Symons, *J. Chem. Soc. (A)*, 1968, 2696.
6.21 B. C. Gilbert and R. O. C. Norman, *Advances in Physical Organic Chemistry*, Ed. V. Gold, 1967, **5,** 53.
6.22 H. Fischer in *Free Radicals*, Vol. II., Ed. J. K. Kochi, Wiley-Interscience, 1973, p. 435.

7.1 P. W. Atkins and M. C. R. Symons, *The Structure of Inorganic Radicals*, Elsevier, Amsterdam, 1967.
7.2 D. R. Smith and J. J. Pieroni, *Canad. J. Chem.*, 1967, **45,** 2723.
7.3 B. L. Bales and L. Kevan, *J. Phys. Chem.*, 1970, **74,** 1098; D. R. Brown and M. C. R. Symons, *J.C.S. Faraday I*, in press.
7.4 W. C. Eastey and W. Weltner, *J. Chem. Phys.*, 1970, **52,** 197.
7.5 R. C. Catton, M. C. R. Symons and H. W. Wardale, *J. Chem. Soc. (A)*, 1969, 2622.
7.6 A. Begum, A. R. Lyons and M. C. R. Symons, *J. Chem. Soc. (A)*, 1971, 2290.
7.7 Z. H. Top, S. Raziel, Z. Luz and B. Silver, *J. Magn. Resonance*, 1973, **12,** 102.
7.8 A. J. Colussi, J. R. Morton, K. F. Preston and R. W. Fessenden, *J. Chem. Phys.*, 1974, **61,** 1247.
7.9 A. J. Colussi, J. R. Morton and K. F. Preston, *J. Chem. Phys.*, 1975, **62,** 2004.

7.10 H. Hasegawa, K. Ohnishi, K. Sogobe and M. Miura, *Mol. Phys.*, 1975, **30,** 1367.
7.11 C. A. McDowell, K. A. R. Mitchell and P. Raghunathan, *J. Chem. Phys.*, 1972, **57,** 1699.
7.12 T. A. Claxton, B. W. Fullam, E. Platt and M. C. R. Symons, *J.C.S. Dalton*, 1975, 1395.
7.13 S. P. Mishra and M. C. R. Symons, *J.C.S. Chem. Comm.*, 1974, 279.
7.14 J. R. Morton, K. F. Preston and J. C. Tait, *J. Chem. Phys.*, 1975, **62,** 2029.
7.15 S. P. Mishra and M. C. R. Symons, *J.C.S. Dalton*, 1976, 1622.

8.1 K.-A. Thuomas and A. Lund, *J. Mag. Res.*, 1976, **22,** 315; S. P. Mishra, G. W. Neilson and M. C. R. Symons, *J.C.S. Faraday II*, 1973, **69,** 1425.
8.2 See, for example, B. C. Gilbert, J. P. Larkin and R. O. C. Norman, *J.C.S. Perkin II*, 1973, 272.
8.3 A. Begum, A. R. Lyons and M. C. R. Symons, *J. Chem. Soc. A*, 1967, 1770.
8.4 A. R. Lyons, G. W. Neilson and M. C. R. Symons, *J.C.S. Faraday II*, 1972, **68,** 1015.
8.5 A. J. Bowles, A. Hudson and R. A. Jackson, *Chem. Phys. Letters*, 1970, **5,** 552.
8.6 M. C. R. Symons, *J.C.S. Faraday II*, 1972, **68,** 1897.
8.7 G. W. Neilson, S. P. Mishra and M. C. R. Symons, *J.C.S. Faraday II*, 1975, **71,** 363.
8.8 R. V. Lloyd, D. E. Wood and M. T. Rogers, *J. Amer. Chem. Soc.*, 1974, **96,** 7130.
8.9 E. D. Sprague and F. Williams, *J. Chem. Phys.*, 1971, **54,** 5425.
8.10 P. J. Krusic and J. K. Kochi, *J. Amer. Chem. Soc.*, 1971, **93,** 846.
8.11 A. R. Lyons and M. C. R. Symons, *J.C.S. Faraday II*, 1972, **68,** 622.
8.12 C. Eaborn, *J. Chem. Soc.*, 1956, 4858; W. Hanstein, J. J. Berwin and T. G. Traylor, *J. Amer. Chem. Soc.*, 1970, **92,** 7476.
8.13 See, for example, E. J. Hart and M. Anbar, *The Hydrated Electron*, Wiley-Interscience, New York, 1970.
8.14 D. J. Nelson, R. L. Petersen and M. C. R. Symons, *J.C.S. Perkin II*, 1977, 2005.
8.15 J. H. Hadley and W. Gordy, *Proc. Natl. Acad. Sci.*, 1974, **71,** 3106; 1975, **72,** 3486.
8.16 B. C. Gilbert, D. K. C. Hodgeman and R. O. C. Norman, *J.C.S. Perkin II*, 1973, 1748.
8.17 B. W. Fullam, S. P. Mishra and M. C. R. Symons, *J.C.S. Dalton*, 1974, 2145.
8.18 I. S. Ginns, S. P. Mishra and M. C. R. Symons, *J.C.S. Dalton*, 1973, 2509.
8.19 A. Begum, A. R. Lyons and M. C. R. Symons, *J. Chem. Soc. A*, 1971, 2290.
8.20 S. A. Fieldhouse, A. R. Lyons, H. C. Starkie and M. C. R. Symons, *J.C.S. Dalton*, 1974, 1966; 1976, 1506.
8.21 R. J. Booth, H. C. Starkie and M. C. R. Symons, *J.C.S. Faraday II*, 1972, **68,** 638.

8.22 B. W. Fullam and M. C. R. Symons, *J.C.S. Dalton*, 1974, 1086.
8.23 M. G. Swanwick and W. A. Waters, *J. Chem. Soc. B*, 1971, 1059.

9.1 M. C. R. Symons, *J. Phys. Chem.*, 1967, **71,** 172.
9.2 H. Sharp and M. C. R. Symons, *Ions and Ion-Pairs in Organic Reactions*, M. Szwarc. Ed., Wiley, New York, N.Y. (1972).
9.3 E. de Boer and E. Sommerdijk, *Ions and Ion-Pairs in Organic Reactions*, M. Szwarc. Ed., Wiley, New York, N.Y. (1972).
9.4 F. C. Adam and S. I. Weissman, *J. Amer. Chem. Soc.*, 1958, **80,** 1518.
9.5 N. Hirota, *J. Amer. Chem. Soc.*, 1968, **90,** 3603; *J. Phys. Chem.*, 1967, **71,** 127.
9.6 T. A. Claxton, J. Oakes and M. C. R. Symons, *Trans. Faraday Soc.*, 1968, **64,** 596.
9.7 T. E. Gough and D. R. Hindle, *Canad. J. Chem.*, 1969, **47,** 1698; 3393.
9.8 T. E. Gough and D. R. Hindle, *Trans. Faraday Soc.*, 1970, **66,** 2420.
9.9 K. S. Chen, T. Takeshita, K. Nakamura and N. Hirota, *J. Phys. Chem.*, 1973, **77,** 708.
9.10 J. Oakes, J. Slater and M. C. R. Symons, *Trans. Faraday Soc.*, 1970, **66,** 546.
9.11 J. Oakes and M. C. R. Symons, *Trans. Faraday Soc.*, 1970, **66,** 1.
9.12 K. Nakamura and N. Hirota, *J. Amer. Chem. Soc.*, 1973, 6919.
9.13 K. S. Chen, S. W. Mao, K. Nakamura and N. Hirota, *J. Amer. Chem. Soc.*, 1971, **93,** 6004.
9.14 N. Hirota and S. I. Weissman, *J. Amer. Chem. Soc.*, 1964, **86,** 2537.
9.15 N. Hirota, *J. Amer. Chem. Soc.*, 1967, **89,** 32.
9.16 G. Goez-Morales and P. Sullivan, *J. Amer. Chem. Soc.*, 1974, **96,** 7232.
9.17 J. Oakes and M. C. R. Symons, *Trans. Faraday Soc.*, 1968, **64,** 2579.
9.18 E. W. Stone and A. H. Maki, *J. Chem. Phys.*, 1962, **36,** 1944.
9.19 E. M. Kosower, *J. Amer. Chem. Soc.*, 1958, **80,** 3253; 3261.
9.20 T. E. Gough and M. C. R. Symons, *Trans. Faraday Soc.*, 1966, **62,** 269.
9.21 D. Jones and M. C. R. Symons, *Trans. Faraday Soc.*, 1971, **67,** 961.
9.22 C. J. W. Gutch, W. A. Waters and M. C. R. Symons, *J. Chem. Soc. (B)*, 1970, 1261.
9.23 C. Jolicoeur and H. L. Friedman, *Ber. Bunsenges Physik. Chem.*, 1971, **75,** 248.
9.24 C. Jolicoeur and H. L. Friedman, *J. Solution Chem.*, 1974, **3,** 15.
9.25 E. Smith and M. C. R. Symons, *J.C.S. Faraday I*, 1976, 2876.
9.26 M. C. R. Symons and M. J. Blandamer, *Hydrogen-Bonded Solvent Systems*, Ed. A. K. Covington and P. Jones, Taylor and Francis, London, 1968.
9.27 D. W. Ovenall and D. H. Whiffen, *Mol. Phys.*, 1961, **4,** 135.
9.28 P. W. Atkins, N. Keen and M. C. R. Symons, *J. Chem. Soc.*, 1962, 2873.
9.29 H. Sharp and M. C. R. Symons, *J. Chem. Soc. (A)*, 1970, 3075.

10.1 See, for example, R. P. Wayne in *Comprehensive Chemical Kinetics*, Vol. II, Ed. C. H. Bamford and C. F. H. Tipper, Elsevier, Amsterdam, 1969.
10.2 See, for example, E. J. Hart and M. Anbar, *The Hydrated Electron*, Wiley-Interscience, New York, 1970.

10.3 G. L. Closs, *Adv. Magn. Resonance*, 1974, **7,** 157.
10.4 See, for example, R. Livingston and H. Zeldes, *J. Chem. Phys.*, 1966, **44,** 1245; 1967, **47,** 1465.
10.5 M. C. R. Symons and M. G. Townsend, *J. Chem. Soc.*, 1959, 263; J. F. Gibson, M. C. R. Symons and M. G. Townsend, *J. Chem. Soc.*, 1959, 269.
10.6 R. W. Fessenden and R. H. Schuler, *Adv. Radiat. Chem.*, 1970, **2,** 56.
10.7 L. B. Knight and K. C. Lin, *J. Chem. Phys.*, 1972, **56,** 6044.
10.8 G. D. Mendenhall, D. Griller and K. U. Ingold, *Chem. in Britain*, 1974, **10,** 248.
10.9 R. O. C. Norman and B. C. Gilbert, *Adv. Phys. Org. Chem.*, 1967, **5,** 53.
10.10 See, for example, *Reactions of Coordinated Ligands*, Adv. in Chem. Series, 1963, Vol. 37.
10.11 H. Fischer, *Z. Naturforch.*, 1964, **19a,** 866.
10.12 M. C. R. Symons, *J. Chem. Soc.*, 1963, 1186.
10.13 A. L. Buley, R. O. C. Norman and R. J. Pritchett, *J. Chem. Soc. (B)*, 1966, 849.
10.14 A. G. Davies, D. Griller and B. P. Roberts, *Angew. Chem. Int. Edn.*, 1971, **10,** 738.
10.15 J. K. Kochi and P. J. Krusic, *Chem. Soc. Spec. Publ.*, 1970, **24,** 147.
10.16 I. S. Ginns, S. P. Mishra and M. C. R. Symons, *J.C.S. Dalton*, 1973, 2509.
10.17 D. Nelson and M. C. R. Symons, *J.C.S. Perkin II*, 1977, 286.
10.18 S. P. Mishra, M. C. R. Symons and B. W. Tattershall, *J.C.S. Faraday I*, 1975, 1772.
10.19 E. D. Sprague and F. Williams, *J. Chem. Phys.*, 1971, **54,** 5425; S. P. Mishra and M. C. R. Symons, *J.C.S. Perkin II*, 1973, 391.
10.20 A. Hasegawa, M. Shiotani and F. Williams, *Discuss. Faraday Soc.*, 1977, No. 63.
10.21 M. C. R. Symons, *J.C.S. Chem. Commun.*, 1977, 403; S. P. Mishra, G. W. Neilson and M. C. R. Symons, *J.C.S. Faraday II*, 1974, **70,** 1280.
10.22 K. U. Ingold in *Free Radicals*, Vol. 1, Ed. J. K. Kochi, Wiley-Interscience, New York, 1973, p. 37.
10.23 R. L. Ward and S. I. Wiseman, *J. Amer. Chem. Soc.*, 1957, **79,** 2086.
10.24 Jagur-Grodzinski and M. Szwarc, in *Ions and Ion Pairs in Organic Reactions*, Ed. M. Szwarc, J. Wiley and Sons, Inc., New York, 1972.
10.25 J. T. Warden and J. R. Bolton, *Accounts of Chem. Res.*, 1974, **7,** 189.
10.26 P. L. Dutton, J. S. Leigh and D. W. Reed, *Biochim. Biophys. Acta*, 1973, **292,** 654.

11.1 C. A. Hutchison and B. W. Mangum, *J. Chem. Phys.*, 1958, **29,** 952; 1961, **34,** 908.
11.2 P. W. Atkins, M. C. R. Symons and P. A. Trevalion, *Proc. Chem. Soc.*, 1963, 222; S. B. Barnes and M. C. R. Symons, *J. Chem. Soc. (A)*, 1966, 66.
11.3 M. C. R. Symons, *Nature*, 1967, **213,** 1226.
11.4 J. A. McRae and M. C. R. Symons, *J. Chem. Soc. (B)*, 1968, 428.
11.5 M. Iwasaki, T. Ichikawa and T. Ohmori, *J. Chem. Phys.*, 1969, **50,** 1991.

11.6 K. D. Bowers, R. A. Kamper and C. D. Lustig, *Proc. Roy. Soc.*, 1959, **251(A),** 565.
11.7 E. Wasserman, *Progr. Phys. Org. Chem.*, 1971, **8,** 319;
11.8 A. Cox, T. J. Kemp, D. R. Payne, M. C. R. Symons, D. M. Allen and P. P. de Moira, *J.C.S. Chem. Commun.*, 1976, 693.
11.9 M. A. El-Sayed, *Accounts Chem. Res.*, 1971, **4,** 23.
11.10 R. H. Clarke and R. M. Hofeldt, *J. Amer. Chem. Soc.*, 1974, **96,** 3005; *J. Chem. Phys.*, 1974, **61,** 4582.
11.11 W. Dietrich, F. Drissler, D. Schmidt and H. C. Wolf, *Z. Naturforsch.*, 1973, **28a,** 284.
11.12 R. R. Lembke, R. F. Ferrante and W. Weltner, *J. Amer. Chem. Soc.*, 1977, **99,** 416.
11.13 *Chemically Induced Magnetic Polarisation*, (eds. A. R. Lepley and G. L. Closs), Wiley, N.Y. 1973. H. R. Ward in *Free Radicals* Vol. 1., (ed. J. K. Kochi), Wiley-Interscience, N.Y. 1973, p. 239. J. K. S. Wan, S.-K. Wong and D. A. Hutchinson, *Accounts Chem. Res.*, 1974, **7,** 58.
11.14 P. W. Atkins and K. A. McLauchlan, in *Chemically Induced Magnetic Polarisation*, Ed. A. R. Lepley and G. L. Closs, Wiley, New York, 1973.

12.1 A. Abragam and B. Bleaney, *Electron Paramagnetic Resonance of Transition Ions*, Oxford University Press, London, 1970.
12.2 J. E. Wertz and J. R. Bolton, *Electron Spin Resonance*, McGraw-Hill, New York, 1972.
12.3 J. L. Petersen and L. F. Dahl, *J. Amer. Chem. Soc.*, 1974, **96,** 2248.
12.4 G. N. Schrauzer and L.-P. Lee, *J. Amer. Chem. Soc.*, 1968, **90,** 6541.
12.5 M. C. R. Symons, *J.C.S. Chem. Commun.*, 1975, 357.
12.6 R. J. Booth and W. C. Lin, *J. Chem. Phys.*, 1974, **61,** 1226.
12.7 B. A. Goodman, D. A. C. McNeil, J. B. Raynor and M. C. R. Symons, *J. Chem. Soc. (A)*, 1966, 1547.
12.8 R. K. Gupta, A. S. Mildvan, T. Yonetani and T. S. Srivastava, *Biochem. Biophys. Res. Commun.*, 1975, **67,** 1005.
12.9 M. C. R. Symons and R. L. Petersen, *Proc. Roy. Soc.* 1978 (in press).
12.10 M. C. R. Symons, D. X. West and J. G. Wilkinson, *J.C.S. Dalton*, 1975, 1696.
12.11 C. M. Guzy, J. B. Raynor and M. C. R. Symons, *J. Chem. Soc. (A)*, 1969, 2299.
12.12 C. T. Delbecq, W. Hayes, M. C. O'Brian and P. H. Yuster, *Proc. Roy. Soc.*, 1963, **A271,** 243.
12.13 A. Carrington, D. J. E. Ingram, K. A. K. Lott, D. S. Schonland and M. C. R. Symons, *Proc. Roy. Soc.*, 1960, **254A,** 101.
12.14 L. E. Mohrmann and B. B. Garrett, *J. Chem. Phys.*, 1970, **52,** 535.
12.15 W. E. Blumberg, in *Magnetic Resonance in Biological Systems*, (eds. A. Ehrenberg, B. E. Malmström and T. Vanngârd), Pergamon Press, London, 1967, p. 119.
12.16 J. F. Gibson, D. J. E. Ingram and D. Schonland, *Disc. Faraday Soc.*, 1958, **26,** 72.
12.17 C. P. Scholes, *Proc. Nat. Acad. Sci.*, 1969, **62,** 428.
12.18 T. D. Smith and J. R. Pilbrow, *Coordination Chem. Rev.*, 1974, **13,** 173.
12.19 L. C. Dickinson, R. H. Dunhill and M. C. R. Symons, *J. Chem. Soc. (A)*, 1970, 922.

12.20 J. G. M. Van Reus and E. de Boer, *Chem. Phys. Letters*, 1975, **31**, 377.

12.21 Unpublished results.

12.22 See, for example, C. L. Hill, J. Renard, R. H. Holm and L. E. Mortenson, *J. Amer. Chem. Soc.*, 1977, **99**, 2549.

12.23 H. Eicher, F. Farak, L. Bogner and K. Gersonde, *Z. Naturforsch*, 1974, **29C**, 683.

12.24 S. Subramaniam and M. C. R. Symons, *J. Chem. Soc. (A)*, 1970, 2367.

12.25 M. C. R. Symons and D. N. Zimmerman, *J.C.S. Dalton*, 1976, 1970.

12.26 P. H. Kasai, S. Evans, A. Hamnett, A. F. Orchard and N. V. Richardson, *J.C.S. Chem. Commun.*, 1974, 921.

12.27 N. Itoh and M. Ikeya, *Solid State Commun.*, 1969, **7**, 355.

12.28 M. C. R. Symons, D. X. West and J. G. Wilkinson, *J.C.S. Chem. Commun.*, 1973, 917.

12.29 H. Taube, *Progr. Inorg. Chem. Radiochem.*, 1959, **1**, 1.

12.30 M. G. Swanwick and W. A. Waters, *J. Chem. Soc. (B)*, 1971, 1059.

12.31 M. C. R. Symons and J. G. Wilkinson, *J.C.S. Faraday II*, 1972, **68**, 1265.

12.32 A. Hudson, M. F. Lappert, P. W. Lednor and B. K. Nicholson, *J.C.S. Chem. Commun.*, 1974, 966.

13.1 L. E. G. Eriksson and A. Ehrenberg, *Acta. Chem. Scand.*, 1964, **18**, 1437; L. E. G. Eriksson, J. S. Hyde and A. Ehrenberg, *Biochim. Biophys. Acta*, 1969, **192**, 211.

13.2 H. M. Swartz, J. R. Bolton and D. C. Borg (Eds.), *Biological Applications of Electron Spin Resonance*, John Wiley and Sons, Inc., New York, N.Y. 1972.

13.3 J. T. Warden and J. R. Bolton, *Accounts of Chem. Res.*, 1974, **7**, 189.

13.4 H. C. Box and E. E. Budzinski, *J. Chem. Phys.*, 1975, **62**, 197.

13.5 M. C. R. Symons and R. L. Petersen, *Proc. Roy. Soc.*, 1978 (in press).

13.6 L. T. Muus and P. W. Atkins (Eds.), *Electron Spin Relaxation in Liquids*, Plenum Press, New York, 1972.

13.7 W. L. Hubbell and H. M. McConnell, *J. Amer. Chem. Soc.*, 1971, **93**, 314.

13.8 B. J. Gaffney and H. M. McConnell, *J. Magn. Res.*, 1974, **16**, 1; A. S. Waggoner, O. H. Griffith and C. R. Christensen, *Proc. Natl. Acad. Sci.*, 1967, **57**, 198.

13.9 D. D. Thomas, J. C. Seidel, J. S. Hyde and J. Gergely, *Proc. Nat. Acad. Sci.*, 1975, **72**, 1729; H. M. McConnell and G. McFarland, *Q. Rev. Biophys.*, 1970, **3**, 91.

13.10 P. Rey and H. M. McConnell, *J. Amer. Chem. Soc.*, 1977, **99**, 1637.

A2.1 A. R. Boate, J. R. Morton and K. F. Preston, *J. Mag. Res.*, 1976, **24**, 259.

A2.2 G. Breit and I. I. Rabi, *Phys. Rev.*, 1931, **38**, 2082.

A2.3 C. P. Byfleet, D. P. Chong, J. A. Hebden and C. A. McDowell, *J. Mag. Res.*, 1972, **2**, 69.

A2.4 C. P. Poole and H. A. Farach, *J. Mag. Res.*, 1971, **4**, 312; 1971, **5**, 305.

A2.5 H. Zeldes, G. T. Trammell and R. Livingston, *J. Chem. Phys.*, 1960, **32**, 618.

A2.6 A. Abragam and B. Bleaney, *Electron Paramagnetic Resonance of Transition Metal Ions,* Oxford University Press, London, 1970.

A2.7 K.-A. Thuomas and A. Lund, *J. Mag. Res.,* 1976, **22,** 315.

A2.8 N. V. Vugman, A. O. Caride and J. Danon, *J. Chem. Phys.,* 1973, **59,** 4418.

A2.9 D. R. Eaton and K. Zaw, *Coord. Chem. Rev.,* 1971, **7,** 197.

A3.1 L. Kevan and L. D. Kispert, *Electron Spin Double Resonance Spectroscopy,* Wiley-Interscience, New York, 1976.

A3.2 G. Feher, *Phys. Rev.,* 1959, **114,** 1219.

Index

Abragam and Bleaney, 123
Alkoxy radical, 153, 154
Alkyl radical, 41–46
Allyl radical, 46
Angular momentum
  orbital, 1, 11
  spin, 1, 11
Aromatic anion radicals, 25, 53–57

Biological systems, 152–157
  Cu(II) proteins, 139
  dinitroxides, 112–114
  high-spin ferric haem, 142–144
  intrinsic radicals, 152, 153
  irradiated materials, 153, 155
  low-spin ferric haem derivatives, 132–134
  metal-ion clusters, 144–148
  oxy-haemoglobin, 135
  photosynthesis, 106, 107
  solvation of nitroxide radicals, 90–92
  spin-label, 155–157
  thymine radical, 154
  vitamin B_{12}, 135
Box, 153
Bridge-head radical, 49, 50.

Carbon-centred α-radicals, 70–73
Carbon-centred β-radicals, 73–77
Corrections, 161–171
  Breit–Rabi, 18, 161–165
  equivalent nuclei, 165
  for small hyperfine coupling, 167–170
  nuclear quadrupole interaction, 18, 70, 71, 170
  nuclear Zeeman interaction, 167–170
Cyclohexadienyl radical, 55, 56

Dipolar interactions
  electron-electron, 109 *et seq.*
  electron-nuclear, 26 *et seq.*
Disulphide anion and cation radicals, 77, 78
Durene cation and anion radicals, 54

ENDOR spectroscopy, 33, 59, 173–175

F-centres, 58, 59, 173
Ferro- and antiferro-magnetism, 145
Fessenden and Schuler, 95
Fischer, 99
Fluorenone anion radical, 87, 88

g-values, 11–15
  for transition-metal complexes, 123 *et seq.*
Generation of radicals, 94–97
  dissociative electron capture, 101
  electron addition, 100, 101
  matrix isolation, 95
  photolysis, 94
  radiolysis, 95
  redox processes, 96, 97
  thermolysis, 95
Gilbert and Norman, 97

Hyperconjugation, 52, 54, 73
Hyperfine interaction
  anisotropic from d-orbitals, 31, 123 *et seq.*
  anisotropic from p-orbitals, 26, 27, 32, 39
  α-proton, 22–25
  β-proton, 43–45
  isotropic, 19 *et seq.*
  signs, 21, 39, 40

Iminoxy radical, 51
Ingold, 103
Ionizing radiation, *see* Radiolyses

Ion-pairs, 82–89
  anions with two binding sites, 84–86
  cation hyperfine coupling, 82–84
  containing radical cations, 88, 89
  solvation of, 86–88
  triple ions and clusters, 88

Jahn–Teller distortion, 136, 137, 139

Kinetic studies, 103–106
  electron-transfer, 103–106
  proton-transfer, 103–104

Line shape, 33
Linewidths, *see* Relaxation processes
Livingston, 94

McConnell, 46
Mechanism, 93–107
  α- and β-scission, 99, 100
  addition, 97–99
  β-elimination, 97–99
  CIDEP, 95, 120–122
  CIDNP, 93, 118–120
  Haber–Weiss mechanism, 96
  induced oxidations, 96
  photolysis, 94
  radiolysis, 93, 95
  thermolysis, 95

Naphthalene anion radical, 105
nitroxide radical, 25, 51, 90–92, 150, 151

Optical detection of resonances, 118
Orbital contributions to hyperfine coupling, 171
Orbital populations, 23, 27, 30, 31, 63
  d-character, 30, 31, 123 *et seq.*
  determination, 15, 177
  p-character, 26–30
  s-character, 20
Organo-nitrogen radicals, 48–50

π-radicals, 46
Phenyl radical, 55–57
Phosphoryl and phosphoranyl radicals, 79, 80,
  99, 100
Photosynthesis, 106–107
Polymer radicals, 97, 98
Powder spectra, 28, 29

Quadroupole effects, 18, 70, 71, 170

Radiolyses, 27, 58, 62, 92
  in biological systems, 153, 155

Relaxation effects, 33–40
  cation transfer, 83–86
  chemical exchange, 36
  electron transfer, 62, 103–106
  group rotations, 38
  nonsecular processes, 33
  proton transfer, 103, 104
  pseudosecular processes, 33, 37
  secular processes, 33
  T_1 and T_2, 33, 34

σ^* radicals, 102
Samples, 158
σ-π overlap, *see* Hyperconjugation
Semiquinone radical, 55, 85, 89, 105, 153
Solvated electrons, 58, 59
Solvation, 89–92
  of aromatic nitro-anions, 90
  of nitroxides, 90–92
  of semiquinones, 89
Spectra, 159–160
Spectra (illustrations)
  ·AH radicals (with proton exchange), 36
  cation of tetraethylenediamine, 49
  ·CH_3, 22
  $CH_3\dot{C}H_2$ (stick diagram), 44
  $CH_3CH_2\dot{C}HOH$, 44
  $Co(CN)_5$—$\dot{N}O(Ph)$, 151
  Cr^{3+}, 141
  Cu^{2+}, 136
  dinitroxide radicals, 113
  H· and D·, 19, 170
  $Me_2\dot{C}CH_2Br$? 75, 76
  $(Me_3C)_2\dot{N}O$, 156
  naphthalene anion, 53
  ·NO_2 (rotating), 37
  O_2^- and O_2^+, 14
  $\dot{C}HCH_2$ / O — O , 38 / CH_2CH_2
  ·PH_4, 67
  polymethylmethacrylate radical, 98
  radical with one $I = \frac{1}{2}$ nucleus, 168
  radical with two equivalent $I = \frac{1}{2}$ nuclei, 167
  radicals with $I = \frac{1}{2}$ (powder), 29
  radicals with $I = 1$ (powder), 28, 35
  $RCH_2\dot{O}$ radicals, 154
  $R_2\dot{C}Br$, 72
  tetracene cations, 174
  thymine radical, 155
  $VO(acac)_2^{2+}$, 129
  VO^{2+} pairs, 146

187

Spectrometers, 4, 5, 13, 158
  ENDOR, 173
Spin density, *see* Orbital populations
Spin exchange, 5
  in diradicals, 112–114
  in metal clusters, 144–148
  triplet-singlet, 112 *et seq.*
Spin-flap satellite lines, 169–170
Spin polarisation, 20, 24, 30, 59, 63, 71
  adjacent atom term, 23
  for transition-metal complexes, 124 *et seq.*
  Q and U Values, 24, 46, 55
Spin trapping, 51, 52
Szwarc, 105

Thiyl radical, 78
Thymine radical, 154
Transition-metal complexes, 123–151
  Ag(0) complexes, 149
  Ag(II) complexes, 136, 139, 147
  clusters in biological systems, 147, 148
  Co(II) complexes, 134, 135
  Cr(III) complexes, 142
  Cu(II) complexes, 136–139
  Cu(II) proteins, 139
  Cu(III) complexes, 147
  Fe(I) complexes, 134
  FeO_4^{2-} and MnO_4^{3-}, 140
  low-spin haem complexes, 132–134
  metal clusters, 144–148
  Met-haemoglobin, 142–144
  Mn(I) complexes, 149, 150
  paramagnetic ligands, 150, 151

$S = \frac{1}{2}$, d^1, 129–132
$S = \frac{1}{2}$, d^5, 132–133
$S = \frac{1}{2}$, d^7, 134, 135
$S = \frac{1}{2}$, d^9, 136–139
$S = 1$, 140
$S = \frac{3}{2}$, 140–142
$S = \frac{5}{2}$, 142–144
  titanium(III) complexes, 131, 132
  unusual valence states, 148–151
  use of metal s and p orbitals, 124, 125
  vanadyl complexes, 129, 130, 145
  vanadyl tartrate, 145, 146
  vitamin B_{12} derivatives, 135
Trapped electrons, *see* Solvated electrons
Triplet state, 1, 4, 5, 16, 108–122
  CH_2 and CR_2, 115
  $C_5H_5^+$, $C_6Cl_6^{2+}$, Ph^+, 117
  ground-state triplets, 115
  of naphthalene, 111, 112
  pair-trapping, 111, 114, 115
  PhN, 116, 117
  SiN_2 and SiCO, 118

Van Vleck, 1, 8
Vinyl radical, 47

Wasserman, 115
Wertz and Bolton, 123
Whiffen, 56

Zero-field splitting, 1, 8, 16, 18
  for $S = \frac{3}{2}$ and $S = \frac{5}{2}$ complexes, 140–144
  of triplet state, 109 *et seq.*

Index of radicals

Organic and Organometallic

(a)
ĊH₃, 22, 41, 45, 48, 64, 65
CH₃ĊH₂, 39, 42, 43, 44
(CH₃)₂ĊH, 39, 42
(CH₃)₃Ċ, 42, 43, 48
CH₂=CH=CH₂, 46
CH₂=ĊH, 47
(CH₃)₂C=C(CH₃)₂⁺, 46
H₂C=Ċ⁻, 47

(b)
H₂ĊOH, 42, 43
HĊ(OH)₂, 43
CH₃ĊHOH, 97
H₂ĊCH₂OH, 97
CH₃CH₂ĊHOH, 44
CH₂=CHĊHOH, 97
H₂C=CCO₂H, 47
HCO; RCO, 48
RO·, 50, 51, 53, 54
R₂ĊOH, 51
R₂ĊO⁻, 51
R₂ĊOH₂⁺, 51
RĊO₂²⁻, 51
RĊO₂H⁻, 51
RĊ(OH)₂, 51
HOĊHCO₂⁻, 103

(c)
H₂C=Ṅ; R₂C=Ṅ, 47, 48
HCN⁻, 48
·NR₃⁺, 48
·NR₂, 48
R₂ĊNR₂ and R₂ĊNR₃⁺, 50
R₂C=ṄO, 51
RNO₂⁻, 55

(CH₃)₂ĊCN, 114
H₂ĊBMe₂, 50

(d)
H₂ĊF, 42, 52, 64, 65
HĊF₂, 42, 52, 64, 65
ĊF₃, 42, 43, 64, 65
H₂ĊCH₂F, 52

(e)
H₂ĊCl, 42
HĊCl₂, 42
CCl₃, 42
Cl₂ĊF, 42
F₃CCl⁻, 102

(f)
Me₂ĊCH₂Br, 75, 76
Me₃C···Br⁻ and Me₃C···I⁻, 76
F₃CBr⁻, 102
F₃CI⁻, 102

(g)
RSSR⁻, RSSR₂, R₂SSR₂⁺, 77, 78
RSS·, 78
RṠO and RṠO₂, 99

(h)
·PR₃⁺, ·P(OR)₃⁺, etc., 79, 80
·P(OR)₄, etc., 80

(i)
CH₃ĊHCH₂AsO₃H, 77
R₂As·, R₃As·⁺, R₄As·⁺, 80

(j)
·SiR₃, ·ALR₃⁻, 80
·PbMe⁺, 81
·HgR, RḢgR⁻, 81

(k)
$C_6H_6^+$, 53
$(C_6H_6)_2^+$, 53
$C_6H_6^-$, 53
$C_6H_5NO_2^-$, 55
$C_6F_6^-$, 56
PhO·, 114

Inorganic

A, 59–60
H·/D·, 20, 169
Ag·, 60, 149
Cd^+, 149
Hg^+, 149
Tl^+, 163

AB, 60–62
CN, 60, 95
NO, 58, 61
N_2^-, 61
F_2^-, 61
Ag_2^+, 61
OH, 62, 96, 97
O_2^+, 13
O_2^-, 14

AB$_2$, 62–64
NO_2, 37, 38, 58, 62, 63
NO_2^{2-}, 31, 62
CO_2^-, 63
SO_2^-, 64
NF_2, 58, 64
O_3^-, 64
N_3^{2-}, 64

FClO, 64
$ClOCl^+$, 64
F_3^{2-}, 64
ClO_2, 58
·HgCN, 95

AB$_3$, 64–66
BH_3^-, 43, 64, 65
NH_3^+, 25, 43, 48, 64, 65
AlH_3^-, 64, 65
SiH_3, 64, 65
PH_3^+, 64, 65, 79
$P_2H_6^+$, 68
CO_3^{3-}, 25, 42, 43, 66
NO_3^{2-}, 31, 66, 105
PO_3^{2-}, 66
SO_3^-, 66
ClO_3, 66
HPO_2^-, 66
HSO_2, 66
NO_3, 66
CO_3^-, 66
ClO_3^{2-}, 66
BrO_3^{2-}, 66
SF_3, 66

AB$_4$, 66–68
PO_4^{2-}, 66
PH_4, 66, 67, 68, 80
PF_4, 67

AB$_5$, 68
PF_5^-, 68
SF_5, 68

AB$_6$, 68
SF_6^-, 68